*Four Colors
Suffice*

Francis Guthrie (1831–99), originator of the four-color problem.

Four Colors Suffice

HOW THE MAP PROBLEM WAS SOLVED

ROBIN WILSON

Princeton University Press
Princeton and Oxford

Published in the United States and the Philippine Islands by
Princeton University Press, 41 William Street, Princeton,
New Jersey 08540

First published 2002 by the Penguin Group, Penguin Books Ltd,
80 Strand, London, WC2R 0RL

Second printing, and first paperback printing, 2005
Paperback ISBN 0-691-12023-4

The cloth edition of this book has been cataloged as follows
Library of Congress Control Number 2002114311
ISBN 0-691-11533-8

This book has been composed in PostScript FF Scala Sans and
PostScript Monotype Fournier

Printed on acid-free paper. ∞

pup.princeton.edu

Printed in the United States of America

10   9   8   7   6   5   4   3   2

In memory of John Fauvel,
who encouraged me to write this book.

Here of a Sunday morning
    My love and I would lie,
And see the coloured counties,
    And hear the larks so high
    About us in the sky.
    A. E. Housman, *A Shropshire Lad*

Suppose there's a brown calf and a big brown dog, and an artist is making a picture of them . . . He has got to paint them so you can tell them apart the minute you look at them, hain't he? Of course. Well then, do you want him to go and paint both of them brown? Certainly you don't. He paints one of them blue, and then you can't make no mistake. It's just the same with maps. That's why they make every state a different color . . .
    Mark Twain, *Tom Sawyer Abroad*

Because the map was printed on a flat surface, only four colors were required to separate each and every state shape from its neighbors. On a sphere, a globe, four colors likewise sufficed. Had the map been printed on a torus – a doughnut shape – seven colors would have been needed to allow for the state-shape distinctions. There are, of course, additional reasons why one seldom encounters a map of the United States on one's doughnut.
    Tom Robbins, *Skinny Legs and All*

Contents

9 *A new dawn breaks* 169

DOUGHNUTS AND TRAFFIC COPS | HEINRICH
HEESCH | WOLFGANG HAKEN | ENTER THE
COMPUTER | COLOURING HORSESHOES

10 *Success! . . .* 190

A HEESCH–HAKEN PARTNERSHIP? | KENNETH
APPEL | GETTING DOWN TO BUSINESS | THE FINAL
ONSLAUGHT | A RACE AGAINST TIME | AFTERMATH

11 *. . . but is it a proof?* 214

COOL REACTION | WHAT IS A PROOF TODAY? |
MEANWHILE . . . | A NEW PROOF | THE FUTURE . . .

Notes and references 229

Chronology of events 245

Glossary 249

Picture credits 255

Index 257

Preface

It is rare for a mathematical problem to catch the attention of the general public. But for a century and a half, the four-colour problem on the colouring of maps has been one of the most famous conundrums in the whole of mathematics, if not *the* most famous. Many thousands of puzzlers, amateur problem-solvers and professional mathematicians have struggled to solve it.

In this book I present the entertaining history of the four-colour problem and its solution. It is a story with many interesting and eccentric characters, including Lewis Carroll, the Bishop of London, a professor of French literature, an April Fool hoaxer, a botanist who loved heather, a mathematician with a passion for golf, a man who set his watch just once a year, a bridegroom who spent his honeymoon colouring maps, and a Californian traffic cop.

After defining the problem, I explain the main ideas of the proof, describe the philosophical problems it has raised, and outline some related colouring problems, ranging from the painting of maps on doughnuts to colouring empires and horseshoes.

For quick reference, a glossary of the technical terms used in the book and a chronology of the main events in the story of the four-colour theorem are included after the notes and references that follow the main text.

I should like to thank the following people, who have made valuable material available to me or have improved the book by their comments: Frank Allaire, Ken Appel, Hans-Günther Bigalke, Norman Biggs, Andrew Bowler, Joy Crispin-Wilson, Robert Edwards, Paul Garcia, Wolfgang Haken, Fred Holroyd, John Koch, Stefan McGrath, Donald MacKenzie, Barbara Maenhaut, David Nelson, Susan Oakes, Toby O'Neil, Adrian Rice, Gerhard Ringel, Ted Swart, Stan Wagon, Ian Wanless, Douglas Woodall and John Woodruff.

The publication of this book coincides with the 150th anniversary of the four-colour problem, and the 25th anniversary of the publication of its proof.

Robin Wilson

May 2002

Four Colors
Suffice

In this map of mainland Britain the counties are coloured with four colours in such a way that neighbouring counties are coloured differently. Can every map be so coloured with only four colours?

The four-colour problem

Before we embark upon our historical journey, there are a number of basic questions to be answered. First and foremost, of course, is this:

What is the four-colour problem?

The four-colour problem is very simply stated, and has to do with the colouring of maps. Naturally, when colouring a map we wish to colour neighbouring countries differently so that we can tell them apart. So how many colours do we need to colour the entire map?

At first sight it seems likely that the more complicated the map, the more colours will be required – but surprisingly this is not so. It seems that at most four colours are needed to colour any map – for example, in the map of Britain shown opposite, the counties have been coloured with just four colours. This, then, is the four-colour problem:

Four-colour problem

Can every map be coloured with at most four colours in such a way that neighbouring countries are coloured differently?

Why is it interesting?

Solving any type of puzzle, such as a jigsaw or crossword puzzle, can be enjoyed purely for relaxation and recreation, and certainly the four-colour problem has provided many hours of enjoyment – and frustration – for many people. At another level, the four-colour problem can be regarded as a challenge. Just as climbing a mountain can present a climber with major physical obstacles to overcome, so this problem – so simple to state, yet apparently so hard to solve – presents mathematicians with an intellectual challenge of enormous complexity.

Is it important?

Rather surprisingly, perhaps, the four-colour problem has been of little importance for mapmakers and cartographers. In an article on the problem's origin written in 1965, the mathematical historian Kenneth May observed:

A sample of atlases in the large collection of the Library of Congress indicates no tendency to minimize the number of colors used. Maps utilizing only four colours are rare, and those that do usually require only three. Books on cartography and the history of mapmaking do not mention the four-color property, though they often discuss various other problems relating to the coloring of maps . . .

The four-color conjecture cannot claim either origin or application
in cartography.

Equally, there seems to have been no interest shown by quilt-
makers, patchworkers, mosaicists and others in restricting the
number of colours of patches or tiles to four.

However, the four-colour problem is more than just a curiosity.
In spite of its recreational nature, the various attempts to solve it
over the years have stimulated the development of much exciting
mathematics with many practical applications to important real-
world problems. Many practical network problems – on road and
rail networks, for example, or communication networks – derive
ultimately from map-colouring problems. Indeed, according to a
recent book in the related area of graph theory (the study of connections between objects), the entire development of the subject
can be traced back to attempts to solve the four-colour problem.
In the theory of computing, recent investigations into algorithms
(step-by-step procedures for solving problems) are similarly linked
to colouring problems. The four-colour problem itself may not be
part of the mathematical mainstream, but the advances it has
inspired are playing an increasingly important role in the evolution
of mathematics.

What is meant by 'solving' it?

To 'prove the four-colour theorem' is to show that four colours are
sufficient to colour all maps – be they geographical maps of the
world, or fictitious ones we may care to invent. If its statement
were false, we would have to demonstrate the fact by presenting a
map that requires five colours or more – just a single map will do.
But if the statement is true, then we must prove it for *all possible*
maps: it is not enough to colour millions or even billions of maps,
because there may still be a map that we have missed whose

arrangement of countries requires five or more colours. In other areas of science, proofs assert that a given hypothesis is over-whelmingly likely given the underlying assumptions and the experiments that have been carried out, but a mathematical proof has to be *absolute* – no exceptions are allowed. To prove the four-colour theorem we must find a general argument that applies to all maps, and discovering such an argument turns out to require the development of a great deal of theoretical machinery.

Who posed it, and how was it solved?

As we shall see, the four-colour problem was first posed by Francis Guthrie around a hundred and fifty years ago, but more than a century of colouring maps and developing the necessary theoretical machinery would pass before it was established with certainty that four colours suffice for all maps. Even then, difficult philosophical questions have remained. The eventual solution, by Wolfgang Haken and Kenneth Appel in 1976, required over a thousand hours of computer time, and was greeted with enthusiasm but also with dismay. In particular, mathematicians continue to argue about whether a problem can be considered solved if its solution cannot be checked directly by hand.

PAINTING BY NUMBERS

Before we begin our historical narrative, we need to explain more clearly what the four-colour problem is, and what our underlying assumptions are. For example, what do we mean by a 'map'?

We may think of a *map* as consisting of a number of *countries* or *regions*. These may be *counties* if part of a British map, or *states* if part of a map of the USA. The boundary of each country is made up of several *boundary lines*, and these boundary lines intersect at various *meeting points*. Two countries with a boundary line in common are called *neighbouring countries* – for example, in the following map the countries *A* and *B* are neighbouring countries.

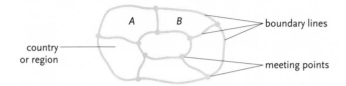

When colouring a map we must always give neighbouring countries different colours. Notice that some maps do need four colours: map (a) below has four mutually neighbouring countries, each meeting the other three; these countries must all be coloured differently, so four colours are required. This can happen in practice: in map (b), *Belgium*, *France*, *Germany* and *Luxembourg* are all neighbouring countries, so the map requires four colours.

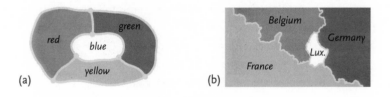

Some maps do not need four colours. For example, in each of the following maps, the outer ring of countries can be coloured with two colours (*red* and *green*) that alternate as we proceed around the ring; the country in the centre must be then assigned another colour (*blue*), so that just three colours are required.

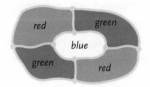 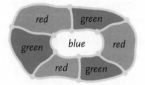

Notice that four colours may be needed even when the map does not contain four mutually neighbouring countries. In the map below, the outer ring of five countries cannot be coloured with two alternating colours, and so requires a third colour. The country in the centre must then be coloured differently from these three, so four colours are required.

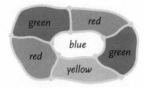

This last situation also arises when we colour the 48 contiguous states of the USA (excluding Alaska and Hawaii). Here, the western state of *Nevada* is surrounded by a ring of five states, *Oregon*, *Idaho*, *Utah*, *Arizona* and *California*: this ring of states requires three colours, and *Nevada* must then be assigned a fourth colour. This colouring with four colours can then be extended to the entire map, as shown opposite.

We can make a couple of further observations about this map. Notice first that at one point of the United States four states meet – *Utah*, *Colorado*, *New Mexico* and *Arizona*. We shall adopt the convention that when two countries meet at a single point, we are allowed to colour them the same – so *Utah* and *New Mexico* may be coloured the same, as may *Colorado* and *Arizona*. This convention is necessary, since otherwise we could construct 'pie maps'

that require as many colours as we choose – for example, the eight-slice pie map below would need eight colours, since all eight slices meet at the centre. With our convention, this map requires only two colours.

Another familiar 'map' that needs only two colours is the chess-board. At each meeting point of four squares we alternate the colours *black* and *white*, producing the usual chessboard colouring (see over).

The other feature of the United States map we need to note is that the state of *Michigan* is in two parts (separated by one of the Great Lakes) that must be assigned the same colour. Although a

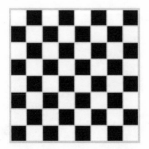

divided country or state may cause no difficulty in particular
situations, such as in the map of the United States, consider the
following map in which the two regions numbered *1* are to be
regarded as parts of the same country. Here, each of the five 'countries' has a boundary line in common with the other four, so five
colours are required:

From now on, we avoid such undesirable situations by requiring
that each country must be in one piece.

Some people also like to include the 'exterior region' in their
colourings. Generally speaking, it is immaterial whether we do so
or not, since we can regard this exterior region as an extra (ring-
shaped) country (see opposite, top).

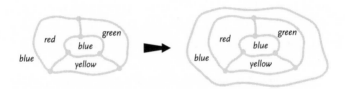

However, a more useful way of thinking about the exterior region is to consider the map as drawn on a globe, rather than on the plane. In this case, the exterior region appears as no different from any other region.

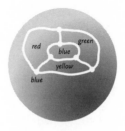

In fact, drawing maps on the plane is equivalent to drawing maps on a globe or, in the language of mathematics, on the surface of a sphere. We can see this from the following illustration, which depicts what is termed a *stereographic projection*, from the

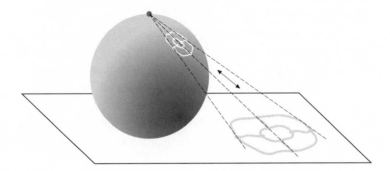

sphere onto the plane on which it rests. Starting with any map on the sphere, we can project it down from the North Pole to give us a map on the plane. Conversely, given any map on the plane, we can project it up onto the sphere.

Notice that these projections do not affect the colouring of the map – if two neighbouring countries are coloured *red* and *green*, then after projection (in either direction) their images will also be *red* and *green*. It follows that we can restate the four-colour problem as a problem about the colouring of maps on a sphere:

Four-colour problem for a sphere

Can every map drawn on the surface of a sphere be coloured with at most four colours in such a way that neighbouring countries are coloured differently?

If we can solve the four-colour problem for maps drawn on a sphere, then we immediately obtain a solution for maps drawn on the plane. Conversely, if we can solve the four-colour problem for maps drawn on the plane, then we immediately obtain a solution for maps drawn on a sphere. So it makes no difference whether we think of our maps as drawn on the plane or on the surface of a sphere and so, for each map, we can choose whether or not to colour the exterior region.

There are some further simplifications which do not materially affect the four-colour problem itself, but which will help to simplify our explanation. It almost goes without saying that every map we consider is in one piece, since we could treat the various pieces as maps in their own right and colour them all separately. Similarly, we can rule out any map that consists of pieces joined together at a single point, since again we could colour the various pieces separately: in particular, we can ignore countries with a single boundary line. Thus, we can ignore any maps like these:

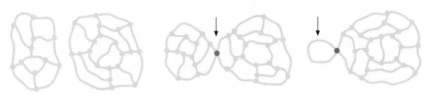

map in two or more pieces   map with pieces joined   one-boundary countries

Finally, we can assume that there are at least three boundary lines at each meeting point. If there were only two boundary lines at a meeting point, then we could simply remove the point without affecting the colouring.

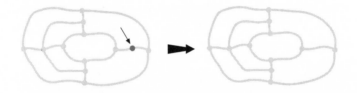

In fact, as we shall see in Chapter 4, when trying to solve the four-colour problem we can always restrict our attention to maps where there are *exactly* three boundary lines at each meeting point, as in the map overleaf. Such maps are very common, and we give them a special name: we call them *cubic maps*. You might like to try to colour this map with four colours. (Can it be coloured with just three?)

TWO EXAMPLES

To end this chapter, here are two practice problems about the colouring of maps on the plane.

Example 1

In the following map, three countries have already been assigned colours. How can we colour the entire map with the four colours *red, blue, green* and *yellow*?

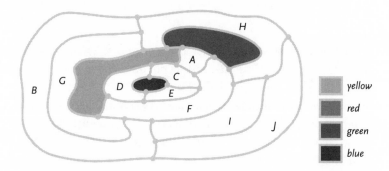

Notice first that country *A* has a boundary line in common with countries coloured *green* and *yellow*, and so must be coloured either *blue* or *red*. Let us look at each of these possibilities in turn.

If country *A* is coloured *blue*, then *F* must be coloured *red* (since it shares a boundary line with countries coloured *blue*, *green* and *yellow*). Country *D* must then be coloured *green*, and *E* must be coloured *yellow*. This gives the following colouring.

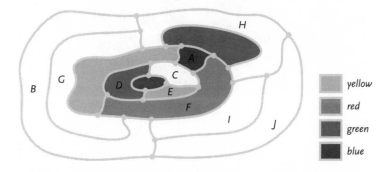

But it is now impossible to colour country *C*, since it has boundary lines in common with countries of all four colours. So *A* cannot be coloured *blue*, and must be coloured *red*.

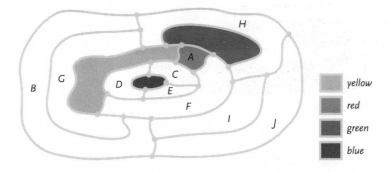

Country *F* must now be coloured *blue*, *C* must be coloured *green*, *D red*, and *E yellow*. We must then colour country *H red*, *G green*, *B yellow*, *I green* and *J blue*. This completes the colouring.

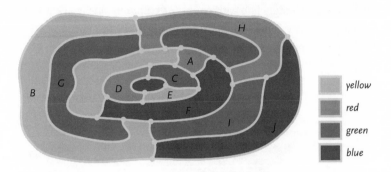

yellow

red

green

blue

Example 2

Our second example first appeared as an April Fool's joke! For several years Martin Gardner wrote a highly successful mathematical column in the pages of *Scientific American*, and many of his columns were later collected in book form. In the 1 April 1975 issue of the magazine he decided to play a trick on his readers by presenting 'Six sensational discoveries that somehow or another have escaped public attention'. They included a chess-playing machine (then considered impossible), a thought experiment that disproved Einstein's theory of special relativity, and the discovery of an ancient manuscript establishing Leonardo da Vinci as the first inventor of the valve flush toilet. The column declared:

> The most sensational of last year's discoveries in pure mathematics was surely the finding of a counterexample to the notorious four-color-map conjecture . . . Last November, William McGregor, a graph theorist of Wappingers Falls, N.Y., constructed a map of 110 regions that cannot be colored with fewer than five colors.

In the July 1975 issue the April Fool hoaxer owned up, reporting that his column had elicited more than a thousand letters and that hundreds of readers had sent in colourings of the map with just four colours. McGregor's map appears below. Can *you* colour it with four colours?

The problem is posed

Unlike many problems in mathematics, the origin of the four-colour problem can be traced precisely – to a letter written in London in 1852. However, for many years it was believed that the problem could be traced back even further – to a lecture given in Germany around 1840. We start our historical narrative by investigating these rival claims and explaining how the confusion arose.

DE MORGAN WRITES A LETTER

On 23 October 1852, Augustus De Morgan, professor of mathematics at University College, London, wrote to his friend Sir William Rowan Hamilton, the distinguished Irish mathematician and physicist. This was nothing unusual. The two men had corresponded for many years, exchanging family news, reporting on the latest scientific gossip in London and Dublin, and sharing bits of mathematical news. Certainly, neither of them could have imagined that the contents of this particular letter would create mathematical history, for it was here that the *four-colour problem* was born.

A student of mine asked
me to day to give him a reason
for a fact which I did not
know was a fact — and do
not yet. He says that, if
a figure be any how divided
and the compartments differently
coloured so that figures with
any portion of common boundary
line are differently coloured
— four colours may be wanted
but not more — The following
is his case in which four
are wanted

A B C & c are
names of
colours

Query cannot a necessity for
five or more be invented

Part of Augustus De Morgan's letter to Sir William Rowan Hamilton,
23 October 1852.

My dear Hamilton, . . .

A student of mine asked me to day to give him a reason for a fact which I did not know was a fact – and do not yet. He says that if a figure be any how divided and the compartments differently coloured so that figures with any portion of common boundary *line* are differently coloured – four colours may be wanted, but not more – the following is his case in which four *are* wanted

A B C D are
names of
colours

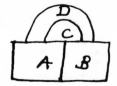

Query cannot a necessity for five or more be invented . . .

What do you say? And has it, if true been noticed? My pupil says he guessed it in colouring a map of England . . . The more I think of it the more evident it seems. If you retort with some very simple case which makes me out a stupid animal, I think I must do as the Sphynx did . . .

Doing as the Sphynx did would have been rather drastic. The Sphynx of ancient mythology was a legendary figure who leapt to her death after Oedipus had correctly solved a difficult riddle she had set him. The riddle was this: What animal walks on four legs in the morning, two at noon, and three in the evening? The answer is Man (as a baby, as an adult, and as an elderly person with a stick).

Years later, the student who had approached De Morgan that fateful day identified himself as Frederick Guthrie, subsequently a physics professor and founder of the Physical Society in London. But it was not Frederick who had coloured the map of England, as he recalled in 1880:

Some thirty years ago, when I was attending Professor De
Morgan's class, my brother, Francis Guthrie, who had recently
ceased to attend them (and who is now professor of mathematics at
the South African University, Cape Town), showed me the fact
that the greatest necessary number of colours to be used in colouring a map so as to avoid identity of colour in lineally contiguous
districts is four. I should not be justified, after this lapse of time, in
trying to give his proof, but the critical diagram was as in the
margin.

With my brother's permission I submitted the theorem to
Professor De Morgan, who expressed himself very pleased with it;
accepted it as new; and, as I am informed by those who subsequently attended his classes, was in the habit of acknowledging
whence he had got his information.

If I remember rightly, the proof which my brother gave did not
seem altogether satisfactory to himself; but I must refer to him
those interested in the subject . . .

Thus it was Frederick Guthrie's elder brother Francis who could
justly claim to have originated the four-colour problem, but the
nature of the 'proof' he gave is not known. Francis Guthrie had
been a former student of De Morgan's at University College,
obtaining a Bachelor of Arts degree there in 1850. Two years later
he took a Bachelor of Laws degree, and was called to the bar in
1857. He had a distinguished career in South Africa, becoming
professor of mathematics at the newly established college at

Graaff-Reinet in the Cape Colony, and later at the South African
College in Cape Town. A well-liked and popular figure, Guthrie
also contributed to botany, which became his chief hobby, and the
plant *Guthriea capensis* and the heather *Erica guthriei* were named
after him. But he never published anything on the colouring of
maps or on the problem that is still sometimes referred to as
Guthrie's problem.

Frederick Guthrie seems to have been the first to observe that
the four-colour problem has no interesting extension to three
dimensions: if we allow three-dimensional 'countries', then we can
construct maps that require as many colours as we wish. An
example, included by him in the note about his brother, involves a
collection of flexible rods (or pieces of coloured wool), all touching
each other. Since each rod must have a different colour from all
those it touches, we need as many colours as there are rods: for
example, the five rods in Guthrie's diagram below require five
colours.

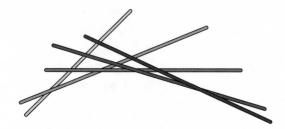

Another three-dimensional example, later described by the Aus-
trian mathematician Heinrich Tietze, involves taking a number of
horizontal bars numbered 1 to n, placing on top of them n vertical
bars also numbered 1 to n, and then joining (as a single 'country')
each pair of horizontal and vertical bars with the same number.
We then obtain n three-dimensional countries, all touching each
other, which therefore require n colours. Here, n can be as large as

we wish: the following pictures show how to construct the five
countries when $n = 5$.

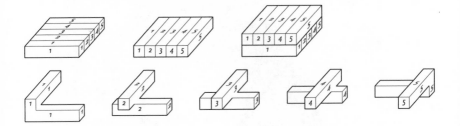

HOTSPUR AND THE *ATHENAEUM*

By 1852 Augustus De Morgan and Sir William Rowan Hamilton
were both well established in their respective careers. De Morgan
had studied at Cambridge University before becoming the first professor of mathematics at the newly founded University College in
London, a position he held for over thirty years. An eccentric and
prolific writer with a style all his own, he is mainly remembered for
his popular book *A Budget of Paradoxes*, for 'De Morgan's laws' in
set theory and for his contributions to mathematical logic. Hamilton was a child prodigy, familiar with Latin, Greek and Hebrew at
the age of five, and speaking Arabic, Sanskrit, Turkish and other
languages by the time he was fourteen. He became Astronomer
Royal of Ireland while still an undergraduate at Trinity College,
Dublin, and held this position until his death in 1865.

As we remarked earlier, De Morgan's 1852 letter to Hamilton
was not an isolated event, for they corresponded regularly for
thirty years. They met only once, around 1830, when they were
introduced to each other by Charles Babbage, whose designs for
the so-called Analytical Engine foreshadowed the invention of the
programmable computer a century later. There was a second

Augustus De Morgan (1806–71)

occasion on which De Morgan and Hamilton were both present, a
Freemasons' dinner in honour of the astronomer and mathematician Sir John Herschel, but the event was so crowded that they
had no chance to speak to each other.

When De Morgan wrote to Hamilton about the map-colour
problem, he doubtless hoped that Hamilton would become interested in it. After all, De Morgan had taken an interest in Hamilton's researches, including Sir William's ground-breaking work on
quaternions. Many mathematical operations are *commutative*,
which is to say that they can be carried out in either order: for
example, the addition and multiplication of ordinary numbers are
commutative ($3 + 4 = 4 + 3$ and $3 \times 4 = 4 \times 3$). However, for Hamilton's quaternions, multiplication is not commutative: his
'numbers' are the sum of four terms $a + bi + cj + dk$ (where a, b, c,
d are numbers and $i^2 = j^2 = k^2 = -1$) that multiply in a non-
commutative way: for example, $i \times j = k$ but $j \times i = -k$, and $k \times i = j$
but $i \times k = -j$.

In the event, De Morgan's letter to Hamilton about the map-
colour problem drew a terse and idiosyncratic reply: 'I am not
likely to attempt your "quaternion" of colours very soon.' Undeterred, De Morgan wrote to other mathematical friends trying to
interest them in the problem. He was fascinated by its intricacies,
and in his original letter to Hamilton he had tried to explain where
the difficulty lies:

As far as I see at this moment, if four *ultimate* compartments have
each boundary line in common with one of the others, three of
them inclose the fourth, and prevent any fifth from connexion with
it. If this be true, four colours will colour any possible map without
any necessity for colour meeting colour except at a point.

Now, it does seem that drawing three compartments with
common boundary *A B C* two and two – you cannot make a fourth
take boundary from all, except inclosing one – But it is tricky work

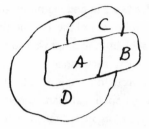

and I am not sure of all convolutions – What do you say? And has
it, if true been noticed?

In this passage De Morgan hits upon the fact that if a map contains four regions, each adjoining the other three, then one of
them must be completely enclosed by the others. He believed,
incorrectly, that this idea lay at the heart of the problem, and it
soon became an obsession of his. Since he could not prove it, he
proposed to assume its truth as an *axiom*, which he defined as 'a
proposition which cannot be made dependent upon obviously
more simple ones'.

In December 1853, De Morgan wrote to the distinguished philosopher William Whewell, Master of Trinity College, Cambridge,
describing his observation as a mathematical axiom that had lain
'wholly dormant' until it arose in connection with the map-colour
problem:

I soon made out the following – which was at first incredible – then
certainly true – then axiomatic – for I cannot make it depend on
anything I see more clearly.

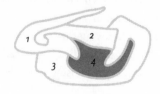

If *four* non-interfering compartments have *each* common boundary *line* with the other three – one at least of the four must be inclosed by the other three – or by fewer . . .

Six months later, in a letter to the Cambridge mathematician Robert Ellis, De Morgan further described it as

an instance of Whewell's views about *latent* axioms, things which at first are not even credible, but which settle down into first principles.

The first known appearance of the four-colour problem in print was also connected with William Whewell. On 14 April 1860, a lengthy unsigned review of Whewell's book *The Philosophy of Discovery, Chapters Historical and Critical* appeared in the *Athenaeum*, a popular literary journal of the time. In his review, the writer outlined the four-colour problem, claiming that the problem was familiar to cartographers. He followed his description with a very obscure passage:

Now, it must have been always known to map-colourers that four different colours are enough. Let the counties come cranking in, as Hotspur says, with as many and as odd convolutions as the designer chooses to give them; let them go in and out and round-about in such a manner that it would be quite absurd in the Queen's writ to tell the sheriff that A. B. could run up and down in his bailiwick: still, four colours will be enough to make all requisite distinction.

This mention of Hotspur refers to a passage in Shakespeare's *King Henry IV, Part 1*, where Hotspur remarks, 'See how this river comes me cranking in . . .'.

The reviewer then asserted that if four areas on a map all have a boundary with the other three, then one area must be surrounded by the others, and this passage clearly identifies De Morgan as the author of the review. De Morgan had indeed written to Whewell on

3 March 1860, thanking him for sending a copy of his book and informing him that he had received a further copy from the *Athenaeum* for review, 'which will go back uncut' (in those days one needed a paper cutter to separate the pages of a book).

As a consequence of De Morgan's *Athenaeum* review, the four-colour problem crossed the Atlantic to the USA. There it was perused by the American mathematician, philosopher and logician Charles Sanders Peirce (pronounced 'purse'), who developed a lifelong interest in the problem. Peirce considered it 'a reproach to logic and to mathematics that no proof had been found of a proposition so simple', and subsequently presented an attempted solution at Harvard University in the presence of his father Benjamin Peirce, the distinguished Harvard professor of mathematics and natural philosophy. As Charles Peirce later wrote:

About 1860 De Morgan in the *Athenaeum*, called attention to the fact that this theorem had never been demonstrated; and I soon after offered to a mathematical society at Harvard University a proof of this proposition extending it to other surfaces for which the numbers of colors are greater. My proof was never printed, but Benjamin Peirce, J. E. Oliver, and Chauncey Wright, who were present, discovered no fallacy in it.

In fact, the seminar probably took place in the late 1860s, but the Peirce manuscripts at Harvard University do not indicate the nature of his solution. His reference to 'extending it to other surfaces' refers to drawing maps on a surface other than a globe (or sphere). For example, suppose that we lived on a world shaped like the surface of an inner tube or ring doughnut – how many colours would we need then? (Mathematicians call such a surface a *torus*.) From his unpublished notes at Harvard, we know that Peirce found a torus map that needed six colours, but in fact we can do even better than this. The following torus map turns out to have seven mutually neighbouring countries and so requires seven

colours (we shall return to the colouring of maps on a torus in
Chapter 7):

Peirce later remarked that the four-colour problem had been useful to him in testing the growth of his logical powers. Indeed, his
researches into mathematical logic included the development of a
'logic of relatives', and in October 1869 he specifically applied this
to map colouring (his approach is outlined in Chapter 5).

In June 1870, at the beginning of an extensive tour of Europe,
Peirce visited the ailing De Morgan in London; it would be fascinating to know whether they discussed the four-colour problem. But
by this time, the problem seems to have been largely forgotten in
England: certainly, there is no evidence that the recipients of De
Morgan's letters were any more interested in the problem than
Hamilton had been on first hearing of it in 1852. On 18 March
1871, Augustus De Morgan died in London, having made little progress with the four-colour problem and unaware that over a century would elapse before a solution was discovered.

MÖBIUS AND THE FIVE PRINCES

As we have seen, the four-colour problem was originated by
Francis Guthrie in 1852. However, it has sometimes been claimed,
incorrectly, that the problem is older than this, dating back to a lecture given by the German mathematician and astronomer August
Ferdinand Möbius around 1840. The *problem of the five princes* that
Möbius posed is superficially similar to the four-colour problem,
and we shall see how they came to be confused.

For many years, Möbius was professor of astronomy in Leipzig
and director of the Leipzig observatory. In mathematics, his name
is associated with the *Möbius function* in the theory of numbers
and with *Möbius transformations* in geometry. But he is best
remembered for the *Möbius strip*, or *Möbius band*, a curious object
constructed from a long rectangular strip of paper by twisting one
end through 180° and then glueing the two ends together, as pic-
tured below. The resulting object has just one side and just one
edge: this means that an ant could travel from any point on it to
any other point without leaving the surface or going over the
boundary edge. It was described by the sixty-eight-year-old

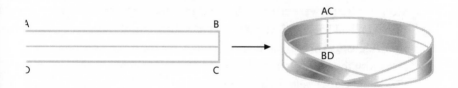

Möbius in late 1858, though it had already been constructed
six months earlier by a professor of optics named Johann
Benedict Listing, who has received little recognition for his prior
discovery.

In one of his lectures on geometry, Möbius asked the following

August Ferdinand Möbius (1790–1868)

question, which had apparently been suggested to him by a Leipzig University friend, the philologist Benjamin Gotthold Weiske, who was greatly interested in mathematics:

Problem of the five princes

There was once a king in India who had a large kingdom and five sons. In his last will, the king said that after his death the sons should divide the kingdom among themselves in such a way that the region belonging to each son should have a borderline (not just a point) in common with the remaining four regions. How should the kingdom be divided?

In the next lecture, Möbius's students admitted that they had tried hard to solve the problem, but without success. Möbius laughed and said he was sorry that they had struggled in vain, as such a division of the kingdom is impossible.

It is easy to see intuitively why Möbius's problem has no solution. Suppose that the regions belonging to the first three sons are called *A*, *B* and *C*. These three regions must all have boundaries in common with one another, as shown opposite in figure (a). The region *D* belonging to the fourth son must now lie completely within the area covered by the regions *A*, *B* and *C*, or completely outside it: these two situations are shown in figures (b) and (c). In each of these situations, it is then impossible to place

the region E belonging to the fifth son so as to have boundaries
with the other four regions, A, B, C and D:

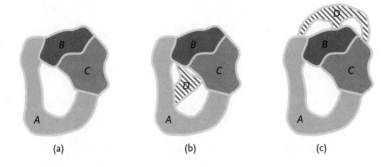

(a)   (b)   (c)

Möbius's problem of the five princes was later extended by Heinrich Tietze, who posed the following related question:

Problem of the five palaces
The king additionally required that each
of his five sons should build a palace in
his region, and that they should link the
five palaces in pairs by roads in such a
way that no two roads cross. How
should the roads be placed?

This problem also has no solution. We can see why, by imitating
the above explanation of the impossibility of solving Möbius's
problem.

Suppose that the palaces belonging to the first three sons are
called A, B and C. These three palaces can be linked by non-
crossing roads, as shown in figure (a) below. The palace D belonging to the fourth son must now lie completely within the area

enclosed by the roads linking A, B and C, or completely outside it: these two situations are shown in figures (b) and (c). In each of these situations, it is then impossible to build the palace E belonging to the fifth son so as to link it by non-crossing roads to the other four palaces A, B, C and D.

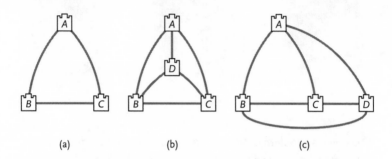

(a)   (b)   (c)

Notice that the solution to either of these problems would have given a solution to the other: if the princes had been able to divide the kingdom into five mutually neighbouring regions, then they would also have been able to build palaces in the regions and construct non-crossing roads joining them. On the other hand, if the princes had been able to build the palaces and the roads joining them, then they would have been able to surround these palaces

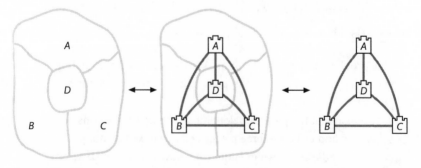

four neighbouring regions   superimpose diagrams   four interconnected palaces by five neighbouring regions. Also, if the king had produced only *four* sons, then the kingdom could easily have been divided, as shown above, and the palaces could have been built and linked by non-crossing roads: with four sons, a solution to either problem yields a solution to the other.

Before leaving Möbius's problem of the five princes, we should note that Heinrich Tietze gave a 'solution' to it. His description continued:

The five brothers sank into despair as it became clear that it was not possible to fulfil the condition of their father's will. Suddenly a wandering wizard, who claimed to possess a solution, was announced . . . We can assume that the wizard was richly rewarded.

The wizard's solution was to connect two of the five regions by a bridge:

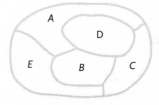

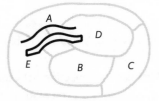

D and E not connected   D and E joined by a bridge

Of course, this is cheating, since we had assumed that we were restricted to drawing the kingdom on a plane, whereas the wizard's solution corresponds to solving Möbius's problem on the surface of a torus:

In fact, for such a problem the king could actually have had up to seven sons: the illustration above shows how to divide a torus into seven neighbouring regions, one for each of the seven sons. But if we are restricted to the plane, then, as we have seen, the maximum number of neighbouring regions is four – in the plane, five mutually neighbouring regions cannot exist.

CONFUSION REIGNS

What is the connection between Möbius's problem of the five princes and the four-colour problem, and why were they confused with each other? Before we answer this, let's have a cup of tea and sort out some logic! I can truthfully say that

if the tea is too hot, then I cannot drink it.

Another way of expressing this is to turn it round, and say that

if I can drink the tea, then it is not too hot.

What I cannot say is that

if the tea is not too hot, then I can drink it,

because there may be many other reasons why I cannot drink it: it may be too strong, or too sweet, or have a dead fly in it.

An arithmetical example of this kind of logic, involving the divisibility of whole numbers, is this:

if a whole number ends with 0, then it is divisible by 5.

For example, the numbers 10, 70 and 530 all end with 0, and all are divisible by 5.

Turning this round, we can deduce that

if a whole number is not divisible by 5, then it cannot end with 0.

For example, 11, 69 and 534 are not divisible by 5, and none of them ends with 0.

But we cannot say that

if a whole number is divisible by 5, then it ends with 0,

because there are many numbers, such as 15, 75 and 535, that are divisible by 5 but do not end with 0.

Logicians like to express such statements using symbols. If we use the letter P to stand for 'the tea is too hot' or 'a whole number ends with 0', and the letter Q to stand for 'I cannot drink it' or 'it is divisible by 5', we can put the first statement in each of the above chains of reasoning into this form:

if P is true, then Q is true – or, equivalently, P implies Q.

Turning this round, we get:

if Q is false, then P is false – or not-Q implies not-P.

But what we cannot say is:

if P is false, then Q is false – or not-P implies not-Q.

Let us now return to Möbius's problem of the five princes. Suppose that the terms of the king's will could be satisfied. Then

the region belonging to each of the five sons would share border-
lines with the regions belonging to the other four sons – that is,
there would be five neighbouring regions, each bordering the
other four. If we wanted to colour these five neighbouring regions
with different colours, we would need five colours (one for each
region). So the four-colour theorem would be false.

The above argument tells us that

> if there is a map with five neighbouring regions, then the four-
> colour theorem is false.

(Here, P is the statement 'there is a map with five neighbouring
regions', and Q is the statement 'the four-colour theorem is
false'.)

Turning this around, as before, we find that

> if the four-colour theorem is true, then there is no map with five
> neighbouring regions.

What we cannot say is that

> if there is no map with five neighbouring regions, then the four-
> colour theorem is true.

So our little argument above, showing the impossibility of solving
Möbius's problem, does not prove the four-colour theorem.

Over the years many people have attempted to prove the four-
colour theorem by showing that no map can have five neighbouring regions. But as we have just seen, this does *not* prove the
required result: their logic is the wrong way round.

One unwary person who fell headlong into this trap was the
German geometer Richard Baltzer. On 12 January 1885 he gave a
lecture to the Leipzig Scientific Society in which he described the
problem of the five princes (which he had discovered among
Möbius's surviving papers) and then explained why there cannot
be five neighbouring regions. Baltzer published the results from

his lecture, wrongly claiming that the four-colour theorem follows immediately from his proof.

Baltzer's published paper was read by Isabel Maddison of Bryn Mawr College, in Philadelphia. In 1897 she wrote a 'Note on the history of the map-colouring problem' in the widely read *American Mathematical Monthly*, mentioning Baltzer's paper and remarking that 'it does not seem to be generally known that Möbius described the question, in a slightly different form, in his lectures in 1840'.

From there, the belief that Möbius was the first to formulate the four-colour problem spread far and wide, and was given credence when various well-known mathematics books, such as Eric Temple Bell's *The Development of Mathematics*, repeated the error. It was not until 1959 that the geometer H. S. M. Coxeter set the story straight, and since then Francis Guthrie has been universally recognized as the true originator of the four-colour problem.

We shall resume our historical development of the four-colour problem in Chapter 4, but first we head back to the eighteenth century to investigate the world of polyhedra.

Euler's famous formula

We begin in Berlin, at the court of Frederick the Great of Prussia. For twenty-five years a regular visitor at Frederick's court was the Swiss mathematician Leonhard Euler (pronounced 'oiler'). Euler had been appointed by Frederick to the Berlin Academy of Sciences in 1741, and became head of the mathematics division.

At first they got on well – Euler even used to send Frederick strawberries from his garden. However, their relationship quickly deteriorated after the Seven Years War of 1756–63, when Russian troops occupied Berlin. Frederick began to take a close interest in the detailed running of the Academy and saw Euler almost every day. Increasingly, Euler came to consider his king as petty and rude, while Frederick, who was a fine composer and performer of music, found the mathematician lacking in social sophistication – rather a bumpkin, in fact. Euler was doubtless relieved to receive an invitation from Empress Catherine the Great in 1766 to return to the St Petersburg Academy of Science in Russia, where he had been head of the mathematics section before taking up his appointment in Berlin. He spent the remainder of his life in St Petersburg.

Euler's abilities at mental arithmetic were legendary. When two students tried to sum a complicated progression and disagreed over the fiftieth decimal place, Euler simply calculated the correct

Leonhard Euler (1707–83)

value in his head. This led the French physicist François Arago to observe, 'He calculated without any apparent effort, just as men breathe, as eagles sustain themselves in the air.'

In spite of his increasing blindness, Euler's mathematical output remained high. He was the most prolific mathematician of all time, producing many hundreds of books and papers amounting to tens of thousands of pages, and the publication of his *Collected Works* is still in progress. He wrote on an enormous range of topics, from the purity of prime numbers, musical harmony and the geometry of the triangle, via calculus and mechanics, to the practical realms of optics, astronomy, acoustics and the sailing of ships. As the applied mathematician Pierre-Simon Laplace would later enthuse to his students,

'Read Euler, read Euler. He is the master of us all.'

EULER WRITES A LETTER

On 14 November 1750, during his time in Berlin, Leonhard Euler wrote a letter to Christian Goldbach, a former colleague at St Petersburg and a talented and enthusiastic mathematician. As De Morgan and Hamilton would do a century later, the two men corresponded for many years, sharing news of the latest mathematical developments. Goldbach is now mainly remembered for posing a conjecture that remains unsolved to this day: that

every even number greater than 2 can be written as the sum of two prime numbers.

For example,

$$10 = 5 + 5, \quad 20 = 13 + 7, \quad 30 = 19 + 11 \quad \text{and} \quad 40 = 23 + 17.$$

Euler's letter was about polyhedra – solid shapes bounded by plane faces, such as a cube (bounded by six squares) or a square pyramid (bounded by a square and four triangles).

The study of polyhedra has a long history, extending at least as far back as the pyramids of ancient Egypt, which date from the third millennium BC. The Greeks were particularly interested in *regular polyhedra*, such as the cube, in which the faces are all regular polygons of the same type (such as squares) and the corners all have the same arrangement of polygons around them. The regular polyhedra, now often called the *Platonic solids*, were discussed by Plato in his dialogue *Timaeus* (*c.* 400 BC). Euclid's *Elements* (*c.* 300 BC), the most widely read mathematics book of all time, shows how to construct them, and concludes with a proof that there can be only five regular polyhedra (see over):

the *tetrahedron*, bounded by four equilateral triangles;

the *cube*, bounded by six squares;

the *octahedron*, bounded by eight equilateral triangles;

the *dodecahedron*, bounded by twelve regular pentagons;

the *icosahedron*, bounded by twenty equilateral triangles.

The Greeks associated these polyhedra with the ancient elements: fire (tetrahedron), earth (cube), air (octahedron) and water (icosahedron); the dodecahedron was associated with the cosmos.

If we now relax the condition that the faces must all be of the same type, but still require the corners to have the same arrangement of regular polygons around them, then we obtain the *semi-regular* (or *Archimedean*) *polyhedra*. There are two infinite families

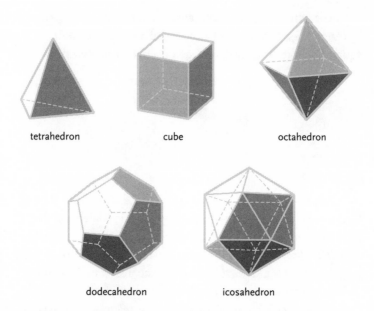

tetrahedron   cube   octahedron

dodecahedron   icosahedron

of these, the prisms and the antiprisms, consisting of a pair of congruent polygons on the top and bottom, with a strip of squares or equilateral triangles around the middle.

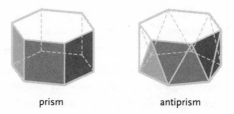

prism   antiprism

There are also thirteen other semi-regular polyhedra, some with wonderful names, such as the snub cube and the great rhombicosi-dodecahedron. Illustrated below are the cuboctahedron (with square and triangular faces), the truncated octahedron (with square and hexagonal faces), the truncated icosahedron (with pentagonal and hexagonal faces) and the great rhombicuboctahedron (with square, hexagonal and octagonal faces).

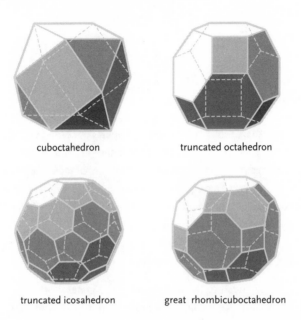

| cuboctahedron | truncated octahedron |

| truncated icosahedron | great rhombicuboctahedron |

These polyhedra are not just mathematical curiosities – they are found widely throughout nature: for example, crystals of iron pyrites occur naturally as cubes, octahedra and dodecahedra, while lead sulphide crystals take the form of cuboctahedra. More recently, certain chemical molecules of carbon atoms, known as *buckminsterfullerene* and nicknamed *buckyballs*, were found to have the form of polyhedra made up from pentagons and hexagons. (The terms buckminsterfullerene and buckyball come from the name of the architect Buckminster Fuller, whose geodesic domes were based on such polyhedra.) The best-known buckyball has the formula C_{60}, with 60 carbon atoms, and takes the form of a truncated icosahedron, resembling the arrangement of pentagons and hexagons on a football.

C_{30} buckyball C_{60} buckyball   football

Although the Greeks and others had been concerned with constructing polyhedra, no one before Euler seems to have had any interest in investigating the possible numbers of faces, edges and vertices (corners) of a polyhedron. It was Euler who first introduced the concepts of *edges* (which he called *acies*) and *vertices* (which he called by their former name of *anguli solidi*, or solid angles).

In his letter to Goldbach, Euler wrote, in the curious mixture of Latin and German that he sometimes employed:

Recently it occurred to me to determine the general properties of solids bounded by plane faces, because there is no doubt that general theorems should be found for them . . .

Letting H be the total number of *hedrae* (faces), s the number of *anguli solidi* (vertices) and A the number of *acies* (edges), Euler then made a number of assertions. They led to a number of propositions, the sixth of which caused him some difficulty:

But I cannot yet give an entirely satisfactory proof of the following proposition:

6. In every solid enclosed by plane faces the aggregate of the number of faces and the number of solid angles exceeds by two the number of edges, or $H + s = A + 2$ . . .

Folgende Proposition aber kann ich nicht recht rigorose demonstriren:

6. In omni solido hedris planis incluso aggregatum ex̄ numero hedrarum et numero angulorum solidorum binario superat numerum acierum, seu est $H + S = A + 2$, seu $H + S = \frac{1}{2}L + 2 = \frac{1}{2}P + 2$.

Part of Euler's letter to Christian Goldbach.

Euler remarked that 'this general result in solid geometry has not previously been noticed by anyone, so far as I am aware'. It is now known as *Euler's polyhedron formula*, or *Euler's formula*, for short, and can be expressed as:

Euler's polyhedron formula

For any polyhedron,

(number of faces) + (number of vertices) = (number of edges) + 2

or, equivalently,

(number of faces) − (number of edges) + (number of vertices) = 2.

Rather than using Euler's choice of notation, we shall write F for the number of faces, V for the number of vertices and E for the number of edges. Euler's polyhedron formula then states that $F - E + V = 2$. We can understand this formula better if we revisit the five regular polyhedra.

For a cube, there are six square faces, twelve edges and eight vertices, so $F = 6$, $E = 12$, $V = 8$, and Euler's formula yields $F - E + V = 6 - 12 + 8 = 2$, as expected. Similarly,

for the tetrahedron, $F = 4$, $E = 6$, $V = 4$,
and $F - E + V = 4 - 6 + 4 = 2$;
for the octahedron, $F = 8$, $E = 12$, $V = 6$,
and $F - E + V = 8 - 12 + 6 = 2$;
for the dodecahedron, $F = 12$, $E = 30$, $V = 20$,
and $F - E + V = 12 - 30 + 20 = 2$;
for the icosahedron, $F = 20$, $E = 30$, $V = 12$,
and $F - E + V = 20 - 30 + 12 = 2$.

Euler's formula also holds for non-regular polyhedra:

for the cuboctahedron, $F = 14$ (8 triangles and
6 squares), $E = 24$, $V = 12$,
and $F - E + V = 14 - 24 + 12 = 2$;
for the truncated octahedron, $F = 14$ (6 squares and
8 hexagons), $E = 36$, $V = 24$,
and $F - E + V = 14 - 36 + 24 = 2$;
for the truncated icosahedron, $F = 32$ (12 pentagons and
20 hexagons), $E = 90$, $V = 60$,
and $F - E + V = 32 - 90 + 60 = 2$;
for the great rhombicuboctahedron, $F = 26$ (12 squares,
8 hexagons and 6 octagons), $E = 72$, $V = 48$,
and $F - E + V = 26 - 72 + 48 = 2$.

Euler's formula has sometimes been wrongly ascribed to the French philosopher René Descartes. Around 1639, Descartes became interested in the total number of angles in all the faces of a polyhedron with F faces and V solid angles (vertices), and obtained the formula $2F + 2V - 4$. A cube, for example, has six square faces, each with four angles, giving a total of 24 angles, and

$$2F + 2V - 4 = (2 \times 6) + (2 \times 8) - 4 = 24.$$

Descartes also observed that each face has as many angles as sides. But each side is common to two faces, so the total number of angles is equal to $2E$. If we equate these two results we obtain $2F + 2V - 4 = 2E$, from which Euler's formula follows – but Descartes himself never made this link, not least because he did not have the appropriate concepts to do so. *The credit for first stating the polyhedron formula thus belongs to Euler.* Descartes's work on polyhedra is now familiar to us through a copy made by Gottfried Wilhelm Leibniz in the 1670s, but this copy did not come to light until 1859, long after Euler's work on the subject had become well established.

Although Leonhard Euler had initially been unable to 'give an entirely satisfactory proof' of his formula, he eventually obtained a 'proof by dissection' in which he successively removed tetrahedron-shaped pieces from the polyhedron in such a way that $F - E + V$ remains unchanged at each stage. Continuing in this way, he eventually reached a single tetrahedron, for which $F - E + V = 2$, as we saw above. Since the value of $F - E + V$ remains the same at each stage of the procedure, it must therefore equal 2 for the polyhedron with which he had started. Euler presented his proof to the St Petersburg Academy of Sciences on 9 September 1751, following it with two substantial papers on polyhedra in the Academy's proceedings, written in 1752 but not published until 1756.

Unfortunately for Euler, it is not at all obvious that his dissection procedure can always be carried out, so his proof must be regarded as deficient. Indeed, the first correct proof was not given until 1794, by the French number-theorist, astronomer and textbook-writer Adrien-Marie Legendre. It appeared in his book *Eléments de géométrie*, and used ideas of angle and area that were more akin to Descartes's approach than to Euler's.

FROM POLYHEDRA TO MAPS

What have all these formulas for polyhedra to do with maps? We can see the connection by projecting the polyhedron from a single point onto a plane surface, such as a table top. To do this, we first 'inflate' the polyhedron out to a sphere, and then project the sphere down onto the plane, as we explained in Chapter 1. The following diagrams show how a cube is projected in this way.

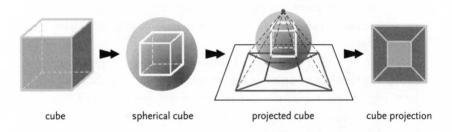

cube   spherical cube   projected cube   cube projection

We can obtain a plane version of any polyhedron in the same way. The following diagrams illustrate a dodecahedron in both its solid and projected versions.

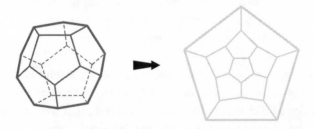

In 1813, Augustin-Louis Cauchy became interested in polyhedra. Cauchy was the leading mathematician in France in the first half of the nineteenth century, obtaining the first rigorous version of the

calculus and helping to initiate an important area of algebra now called group theory. He probably has more results and concepts named after him than does any other mathematician.

Cauchy proved Euler's formula by first projecting the polyhedron onto a plane surface, as just described. The formula still holds for these plane drawings, provided that we are careful how we deal with the exterior region. Cauchy preferred to ignore this exterior region (which appeared when we flattened the polyhedron), so that he had one region fewer than before. His statement of Euler's formula thus became $F - E + V = 1$; for example, the flattened cube has 5 'interior' regions, 12 edges and 8 vertices, so $F - E + V = 5 - 12 + 8 = 1$. However, if we include the exterior region, as we usually choose to when dealing with projections of polyhedra, then $F = 6$, and $F - E + V = 6 - 12 + 8 = 2$, as before.

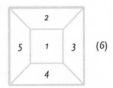

(6)

And this is where the connection comes in. We can also think of such plane drawings as maps – the faces correspond to the countries, the edges correspond to the boundary lines, and the vertices correspond to the meeting points.

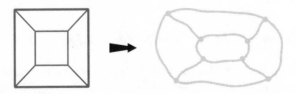

Euler's formula then takes the following form:

Euler's formula for maps

If we include the exterior region, then (no. of countries) − (no. of boundary lines) + (no. of meeting points) = 2.

Why is Euler's formula true? One explanation has some features in common with Euler's attempted proof by dissection. We draw a picture of the polyhedron (or map) in the plane and then remove its edges one at a time, always keeping the diagram in one piece, and see what happens to $F - E + V$ at each stage.

How do we remove the edges? There are two possibilities:

1 The removed edge may separate two different faces

Removing such an edge decreases the number of edges by 1, leaves the number of vertices the same, and decreases the number of faces by 1 (since the two faces become merged into one). Thus, E and F are decreased by 1 and V remains unchanged, so $F - E + V$ remains the same.

2 The removed edge is a 'hanging edge'

Removing such an edge (and the 'hanging vertex' attached to it) decreases the number of edges by 1, decreases the number of vertices by 1, and leaves the number of faces the same. Thus, E and V

are decreased by 1 and F remains unchanged, and F − E + V again remains the same.

The following diagrams illustrate this procedure when we successively remove edges from a tetrahedron. Notice how F − E + V remains unchanged throughout the process (in each case, we include the exterior region).

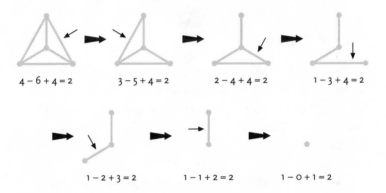

$4 - 6 + 4 = 2$ $3 - 5 + 4 = 2$ $2 - 4 + 4 = 2$ $1 - 3 + 4 = 2$

$1 - 2 + 3 = 2$ $1 - 1 + 2 = 2$ $1 - 0 + 1 = 2$

Whatever polyhedron or map we start with, we always end up with a single vertex, for which the value of F − E + V is 2. Since this expression remains unchanged throughout the process, we deduce that F − E + V = 2 for the original polyhedron. This explains Euler's formula.

Around the same time that Cauchy was publishing his paper, a Swiss mathematician named Simon-Antoine-Jean Lhuilier showed that Euler's formula can fail to hold for polyhedra with certain strange properties. Lhuilier listed three types of strange polyhedra, of which we shall consider just one. If we have a polyhedron with one or more tunnels bored through it, we obtain a version of Euler's formula in which $F - E + V$ is no longer equal to 2. For example, Lhuilier considered a cube with a single tunnel through it, and added extra edges so that each face is surrounded by a polygon. The resulting 'polyhedron' has 16 faces, 32 edges and 16 vertices, so that $F - E + V = 16 - 32 + 16 = 0$:

It turns out that $F - E + V = 0$ for any polyhedron with a single tunnel through it. More generally, Lhuilier proved that whenever we bore another tunnel through the solid, we lower the right-hand side of Euler's formula by 2. Thus, for a polyhedron with two tunnels through it, $F - E + V = -2$, and in general, for a polyhedron with k tunnels, $F - E + V = 2 - 2k$.

In the rest of this chapter we look at two applications of Euler's formula. If you are interested mainly in the historical development of the subject you can skip the mathematical details, but do note the main results.

ONLY FIVE NEIGHBOURS

All the polyhedra we have met in this chapter contain at least one triangle, square or pentagon, and this is indeed true for all polyhedra. The corresponding result for maps drawn on the plane or sphere is this:

'Only five neighbours' theorem
Every map has at least one country with five or fewer neighbours.

This result is very important – indeed, it is central to any proof of the four-colour theorem. Since we have excluded 1-sided countries, it tells us that every map must contain at least one 2-sided country (a *digon*), 3-sided country (a *triangle*), 4-sided country (a *square*) or 5-sided region (a *pentagon*):

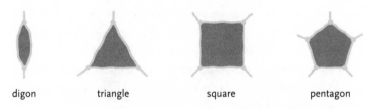

digon   triangle   square   pentagon

To prove the 'only five neighbours' theorem, we consider a map with F countries, E boundary lines and v meeting points, and use Euler's formula. From our discussion of maps in Chapter 1 we can assume that in our map there are at least three boundary lines at each meeting point.

Next, we count the boundary lines emerging from all these meeting points. We appear to get at least $3v$ boundary lines in total, because there are at least three boundary lines at each of the v meeting points. But each boundary line has a meeting point at each end, and is thus counted twice – so we must divide by 2. So E, the total number of boundary lines, is at least $\frac{3}{2}v$. In symbols, $E \geq \frac{3}{2}v$, which we can turn round and rewrite as $v \leq \frac{2}{3}E$.

We prove that the map has at least one country with five or fewer neighbours by assuming the opposite – that our map has no such country – and showing that this assumption leads to an absurdity. It follows from this assumption that each country is sur-

rounded by at least six neighbours. So, counting up the boundary
lines around all the countries we appear to get at least $6F$ boundary lines in total, because each of the F countries is surrounded
by at least six boundary lines. But again, each line has been
counted twice, because it has a country on each side, so we must
divide by 2. So E is at least $\frac{6}{2}F$ (which is $3F$). In symbols, $E \geq 3F$,
which we can rewrite as $F \leq \frac{1}{3}E$.

   We now put these two inequalities $F \leq \frac{1}{3}E$ and $V \leq \frac{2}{3}E$ into
Euler's formula, and we get

   $F - E + V \leq \frac{1}{3}E - E + \frac{2}{3}E$, which is 0.

But by Euler's formula, $F - E + V$ is equal to 2, so we have managed to prove the remarkable result that $2 = 0$. This is clearly false.
Our error has arisen because we assumed that each country has at
least six neighbours – so this assumption is unjustified. It follows
that at least one country has five or fewer neighbours – which is
what we set out to prove.

A COUNTING FORMULA

Our second application of Euler's formula is to obtain a result
which has interesting consequences for polyhedra, and which we
shall need in Chapter 8 – it is called the *counting formula*. For simplicity (and because this is the case that interests us), we limit ourselves to cubic maps, maps in which there are exactly three
boundary lines at each meeting point.

   Suppose that the map has c_2 two-sided countries (digons),
c_3 three-sided countries (triangles), c_4 four-sided countries
(squares), and so on. Then F, the total number of countries
(including the exterior region), is the sum of all of these numbers:

   $F = c_2 + c_3 + c_4 + c_5 + c_6 + c_7 + \dots$ (1)

We next count up all the boundary lines in our map, in terms of these numbers:

since each digon has two boundary lines, the c_2 digons are surrounded by a total of $2c_2$ lines;

since each triangle has three boundary lines, the c_3 triangles are surrounded by a total of $3c_3$ lines;

since each square has four boundary lines, the c_4 squares are surrounded by a total of $4c_4$ lines;

and so on.

Adding up all these numbers, we appear to get E, the total number of boundary lines. In fact, we get $2E$, twice the total number of boundary lines, because each boundary line lies on the boundaries of two countries – that is,

$$2E = 2c_2 + 3c_3 + 4c_4 + 5c_5 + 6c_6 + 7c_7 + \ldots,$$

which we can rewrite as

$$E = c_2 + \tfrac{3}{2}c_3 + 2c_4 + \tfrac{5}{2}c_5 + 3c_6 + \tfrac{7}{2}c_7 + \ldots \tag{2}$$

Similarly, we can find the number of meeting points in our map. Since it is a cubic map, there are exactly three boundary lines at each of the v meeting points, giving a total of $3v$ lines altogether. But since each line joins two of the points, we have counted each line twice. So $3v = 2E$, and we can use the above expression for $2E$ to get

$$3v = 2c_2 + 3c_3 + 4c_4 + 5c_5 + 6c_6 + 7c_7 + \ldots,$$

which we can rewrite as

$$v = \tfrac{2}{3}c_2 + c_3 + \tfrac{4}{3}c_4 + \tfrac{5}{3}c_5 + 2c_6 + \tfrac{7}{3}c_7 + \ldots \tag{3}$$

Thus we have expressed F, E and v in terms of the numbers c_2, c_3, $c_4, \ldots$

What happens when we substitute the three numbered equations into Euler's formula? We have:

$$2 = F - E + V$$
$$= (c_2 + c_3 + c_4 + c_5 + c_6 + c_7 + \ldots)$$
$$- (c_2 + \tfrac{3}{2}c_3 + 2c_4 + \tfrac{5}{2}c_5 + 3c_6 + \tfrac{7}{2}c_7 + \ldots)$$
$$+ (\tfrac{2}{3}c_2 + c_3 + \tfrac{4}{3}c_4 + \tfrac{5}{3}c_5 + 2c_6 + \tfrac{7}{3}c_7 + \ldots).$$

Rearranging this gives

$$2 = c_2(1 - 1 + \tfrac{2}{3}) + c_3(1 - \tfrac{3}{2} + 1) + c_4(1 - 2 + \tfrac{4}{3}) + c_5(1 - \tfrac{5}{2} + \tfrac{5}{3})$$
$$+ c_6(1 - 3 + 2) + c_7(1 - \tfrac{7}{2} + \tfrac{7}{3}) + \ldots$$
$$= \tfrac{2}{3}c_2 + \tfrac{1}{2}c_3 + \tfrac{1}{3}c_4 + \tfrac{1}{6}c_5 + 0c_6 - \tfrac{1}{6}c_7 - \ldots$$

Finally, multiplying by 6 to eliminate fractions, we obtain our required counting formula:

Counting formula for cubic maps

$$4c_2 + 3c_3 + 2c_4 + c_5 - c_7 - 2c_8 - 3c_9 - \ldots = 12.$$

Notice that the coefficients (the numbers 4, 3, 2, . . . in front of the c's) successively decrease by 1, and that c_6 (the number of six-sided countries) does not appear, since its coefficient is 0.

As a simple illustration of the counting formula, consider this cubic map:

This map has nine 4-sided countries and two 9-sided countries (including the exterior one), so $c_4 = 9$ and $c_9 = 2$; all other terms are 0. The counting formula then reduces to

$$2c_4 - 3c_9 = (2 \times 9) - (3 \times 2),$$

which is 12, as expected.

Notice that, in the counting formula, not all of the numbers c_2, c_3, c_4 and c_5 can be 0, since otherwise the first four terms on the left-hand side of the equation would disappear, making the left-hand side a negative quantity. But that is impossible, since the right-hand side is 12, a positive number. It follows that at least one of the numbers c_2, c_3, c_4 and c_5 must be positive, and so there must be at least one digon, triangle, square or pentagon in the map. For cubic maps this gives us an alternative proof of the 'only five neighbours' theorem, that every map has at least one country with five or fewer neighbours.

Also, if our map contains no digons, triangles or squares, the counting formula reduces to

$$c_5 - c_7 - 2c_8 - 3c_9 - 4c_{10} - \ldots = 12.$$

Since the only positive term on the left-hand side is the first one, c_5 must be at least 12. It follows that

every cubic map that contains no digons, triangles or squares must contain at least twelve pentagons.

We shall use this fact in Chapter 8.

Finally, here are three results on polyhedra that follow from the above counting formula for cubic maps. These results hold for any *cubic polyhedron* – a polyhedron in which exactly three faces meet at each vertex.

If all the faces of a cubic polyhedron are pentagons and hexagons, then there are exactly twelve pentagons.

This is because the only non-zero terms are c_5 and c_6, and the counting formula reduces to $c_5 = 12$: an example of such a polyhedron is a *truncated icosahedron* – so all buckyballs and footballs must have twelve pentagons!

If all the faces of a cubic polyhedron are squares and hexagons, then there are exactly six squares.

This is because the only non-zero terms are c_4 and c_6, and the counting formula reduces to $2c_4 = 12$, giving $c_4 = 6$: an example of such a polyhedron is a *truncated octahedron*.

If all the faces of a cubic polyhedron are squares, hexagons and octagons, then the number of squares exceeds the number of octagons by 6.

This is because the only non-zero terms are c_4, c_6 and c_8, and the counting formula reduces to $2c_4 - 2c_8 = 12$, giving $c_4 - c_8 = 6$: an example of such a polyhedron is a *great rhombicuboctahedron*.

Cayley revives the problem . . .

After Augustus De Morgan's death in 1871, the map-colouring problem lay dormant. Although Charles Sanders Peirce in the United States continued to try to solve it, none of De Morgan's British friends made any mention of it. Yet the problem was not completely forgotten: it certainly continued to languish in the mind of Arthur Cayley of Cambridge University.

Arthur Cayley, a cousin of the inventor and pioneering aviator Sir George Cayley, was an outstanding student at Trinity College, Cambridge. He graduated as Senior Wrangler (best of the year) in 1842, and in October of that year was elected a Fellow of Trinity, the youngest person appointed to a Cambridge Fellowship in the nineteenth century. However, the rules of the university at that time required its fellows to enter holy orders within seven years of receiving their MA degree. Unwilling to do so, Cayley left Cambridge to enter the legal profession, studying at Lincoln's Inn in London.

Cayley was called to the Bar in 1849, and practised law very successfully for fourteen years. During this period he continued his mathematical researches, publishing no fewer than three hundred papers, many of which contained his best and most original work. In particular, he was one of the founding fathers of matrix algebra, and wrote the first important paper on the subject in 1858. Much

Arthur Cayley (1821–95)

of his work on algebra in the early 1850s was with the brilliant but unpredictable James Joseph Sylvester, who had been unable to secure a mathematical position in England and whom we shall meet again in the next chapter.

In 1863 Cayley was elected to the newly founded Sadleirian Chair of Pure Mathematics at Cambridge, obtaining dispensation from the religious requirements. He was elected to an Honorary Fellowship at Trinity College, where he remained for the rest of his life.

CAYLEY'S QUERY

On 13 June 1878, Arthur Cayley attended a meeting of the London Mathematical Society (LMS). This august body had been founded at University College, London, in 1865, with Augustus De Morgan as its first president, succeeded in turn by Sylvester and Cayley. By 1878 the number of members had increased from an initial figure of 27 to about 150.

At this particular meeting a former president, Professor Henry Smith of Oxford University, was in the chair, and papers were read on topics ranging from 'the fluxure of spaces' to 'finding differential resolvents of algebraical equations'. As recorded in the Society's *Proceedings*, Cayley then raised a query:

Questions were asked by Prof. Cayley, F.R.S. (Has a solution been given of the statement that in colouring a map of a country, divided into counties, only four colours are required, so that no two adjacent counties should be painted in the same colour?)

On 11 July, Cayley's query was repeated in a report of the meeting in *Nature*.

For almost a century, from then until 1976 when De Morgan's *Athenaeum* review of Whewell's *The Philosophy of Discovery* was located (see Chapter 2), these reports were believed to be the earliest printed references to the four-colour problem. One hundred years after that LMS meeting, on 13 June 1978, I commemorated the centenary of Cayley's query with an item about the problem on the BBC's morning radio *Today* programme, thereby enhancing breakfast time for several million people. (An earlier suggestion for commemorating the centenary with a television documentary was turned down by the BBC *Horizon* producers as being of 'insufficient social relevance'; the following week, *Horizon* broadcast a documentary on the sex life of alligators in South America!)

Cayley was genuinely interested in the four-colour problem, and soon after presenting his query he published a short note on the subject – not in an established mathematical journal, but in a brand new geographical one. In the April 1879 issue of the *Proceedings of the Royal Geographical Society*, he admitted that 'I have not succeeded in obtaining a general proof: and it is worth while to explain wherein the difficulty consists.'

Cayley began by mentioning that the result is 'mentioned somewhere by the late Professor de Morgan, who refers to it as a theorem known to map-makers': we now know that the 'somewhere' was the *Athenaeum* review of 1860. Cayley then stated the four-colour problem and gave the usual four-country example to show that four colours may be needed. He next observed that

if a system of an arbitrary number n of areas has already been coloured with four colours, and if an $(n + 1)$th area is then added, it by no means follows that we can colour the new area with the same four colours, without altering the original colouring.

For example, he said, if the original colouring has all four colours present on the boundary, and if the new area surrounds them all, then there is no colour available for the new area and some recolouring is necessary.

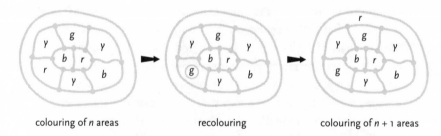

colouring of *n* areas   recolouring   colouring of *n* + 1 areas

Cayley then made the useful observation that, when trying to prove that four colours are always sufficient, one can impose more stringent conditions on our maps. In particular, we can restrict our attention to cubic maps, ones in which there are exactly three countries at each meeting point. To see why this is true, suppose that there are more than three countries at some meeting points in the map. Over each such point we could stick a small circular 'patch', thereby producing a new map in which there are exactly three countries at each meeting point. If we can colour this new map with four colours, as assumed, then we easily obtain a colouring of our original map by shrinking each patch down to a point.

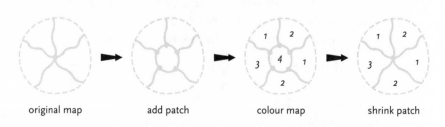

original map   add patch   colour map   shrink patch

In a similar way, we can impose other restrictive conditions – for example, as Cayley observed, if we can colour all maps with four colours, then we can do so in such a way that only three colours appear on the exterior boundary of the map. This is because we can always add a new ring-shaped country that surrounds the map. Colouring this new map with four colours immediately gives us a colouring of the original map with only three colours on the exterior boundary.

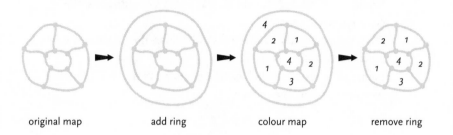

original map   add ring   colour map   remove ring

KNOCKING DOWN DOMINOES

Cayley's remark about taking a coloured map with n countries, adding an $(n + 1)$th country and then trying to extend the colouring to this new country, leads to a method for tackling the problem. It is known to mathematicians as the *method of mathematical induction*, and it can be traced back at least as far as the fourteenth-century French mathematician Levi ben Gerson, inventor of the cross-staff for navigation, who used mathematical induction to prove various results involving permutations and combinations.

In the context of the four-colour problem, the method of mathematical induction takes the following form. Suppose that we can prove, for an arbitrary number n, that

if all maps with n countries can be coloured with four colours, then so can all maps with $n + 1$ countries.

We certainly know that all maps with up to four countries can be coloured with four colours, and so:

taking $n = 4$, we deduce that all maps with 5 countries can be coloured with four colours;
then, taking $n = 5$, we deduce that all maps with 6 countries can be coloured with four colours;
then, taking $n = 6$, we deduce that all maps with 7 countries can be coloured with four colours;
and so on.

Continuing in this way, we deduce that *all* maps can be coloured with four colours.

We can think of a proof by induction as knocking down an unending line of dominoes. We start by knocking down the first domino (or, here, the first four), which in turn knocks down the next one. By our assumption, each domino knocks down the next one – the nth knocks down the $(n+1)$th – so that eventually all the dominoes must fall.

But how can we extend a colouring of a map with n countries to a colouring of a map with $n + 1$ countries? In some instances this is easy: the following colouring extends directly, without any recolouring, and we can colour the extra country *red* (opposite, top).

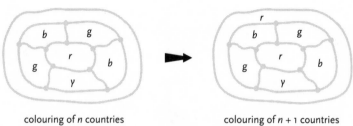

colouring of *n* countries   colouring of *n* + 1 countries

Sometimes we cannot extend the colouring directly, as happens in the next example below where four different colours appear on the exterior boundary. But if, on the left of the map, we recolour the *red* country *green* and the *green* country *red*, then we can colour the extra country *red*.

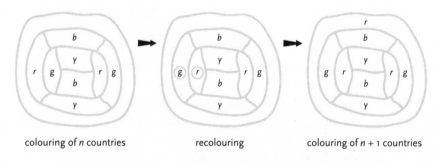

colouring of *n* countries   recolouring   colouring of *n* + 1 countries

In such simple cases we can extend the given colouring to an extra country without much difficulty, but as our maps become ever more complicated we may need to do a substantial amount of recolouring before we can colour the extra country. You might like to try to extend the colouring of the map overleaf to the uncoloured country in the middle. (We shall meet this example again in Chapter 7.)

It is clear that a general method for extending a colouring will be hard to find, and therein lies the immense difficulty of the four-colour problem.

MINIMAL CRIMINALS

There is an alternative but related approach to solving the four-colour problem. Imagine that the four-colour theorem is false, and that there exist some maps that cannot be coloured with four colours. Among these particular maps that need five colours or more, there will be some with the smallest possible number of countries. We call such maps *minimal counter-examples*, or *minimal criminals* – they exhibit 'criminal' behaviour in that they cannot be coloured with four colours, and they are 'minimal' because, subject to this condition, they have as few countries as possible. It follows that

> a minimal criminal *cannot* be coloured with four colours, but
> any map with fewer countries *can* be coloured with four colours.

To prove that four colours suffice, we must prove that minimal criminals cannot exist, and we try to do this by finding ever more restrictive conditions to impose on them. In fact, we try to make life so difficult for them that they cannot exist!

As an example, we can easily show that a minimal criminal cannot contain a two-sided country (a digon). For, suppose that there is a minimal criminal that contains a digon, as shown below. If we remove a boundary line from the digon, merging the digon with one of its neighbours, we obtain a new map with fewer countries. By our assumption, we can colour this new map with four colours.

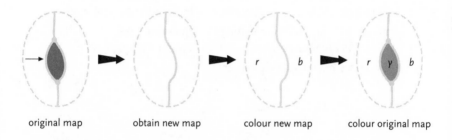

original map   obtain new map   colour new map   colour original map

We now reinstate the digon by replacing the removed line. Since four colours are available, and since the countries next to the digon use only two of them, there must be a spare colour for the digon. Thus, we can colour the minimal criminal with four colours, which contradicts our assumption. This shows that a minimal criminal cannot contain a digon.

In a similar way, we can show that a minimal criminal cannot contain a three-sided country (a triangle). Suppose that it did, as shown overleaf. We now remove a boundary line, merging the triangle with one of its neighbours. We thus obtain a map with fewer countries which we can colour with four colours, as before.

We now reinstate the triangle. Since the countries next to the triangle use only three of the colours, we can use the fourth to colour the triangle. We have thus coloured the minimal criminal with four colours, contradicting our assumption. So a minimal criminal cannot contain a triangle.

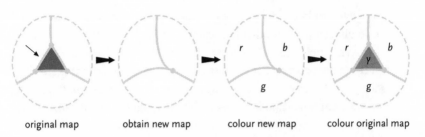

original map   obtain new map   colour new map   colour original map

What happens if we try to extend these ideas to a minimal criminal that contains a four-sided country (a square)? As before, we
remove a boundary line, merging the square with one of its neighbours and obtaining a map with fewer countries. Again, we can
colour this new map with four colours.

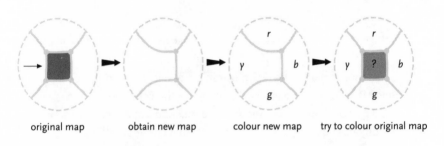

original map   obtain new map   colour new map   try to colour original map

But when we reinstate the square, the countries next to it may
already use all four colours, in which case no spare colour is available for colouring the square and the proof cannot proceed as
before.

A similar thing happens if we try to extend these ideas to a
minimal criminal that contains a five-sided country (a pentagon).
As before, we remove a boundary line, merging the pentagon with
one of its neighbours and obtaining a map with fewer countries.
As in all the previous cases, this new map can be coloured with
four colours.

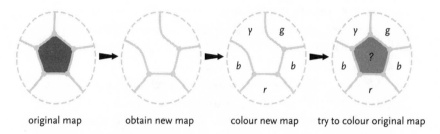

original map   obtain new map   colour new map   try to colour original map

When we reinstate the pentagon, the countries next to the pentagon may already use all four colours, leaving no spare colour for the pentagon. Again, the proof cannot proceed.

In the next chapter we shall see how Alfred Kempe overcame the difficulty when the minimal criminal contains a square, using his *method of Kempe chains*, and then tried to extend his ideas to a minimal criminal that contains a pentagon.

THE SIX-COLOUR THEOREM

The idea of a minimal criminal can be used to prove that every map can be coloured with six colours. Although rather weaker than the four-colour theorem, it is still a remarkable result.

Six-colour theorem
Every map can be coloured with six colours in such a way that neighbouring countries are coloured differently.

To prove the six-colour theorem, we first assume that it is false, and then look for a contradiction. By our assumption there are maps that cannot be coloured with six colours. Among these maps that require seven colours or more, we consider a minimal criminal, one with the smallest possible number of countries. This map *cannot* be coloured with six colours, but any map with fewer countries *can* be coloured with six colours.

We now use the 'only five neighbours' theorem, from Chapter 3. This theorem tells us that our map must contain a country *C* with five or fewer neighbours, as shown below. We now remove a boundary line from the country *C*, merging the country with one of its former neighbours. We then obtain a map with fewer countries, which by our assumption we can colour with six colours, *red*, *blue*, *green*, *yellow*, *purple* and *white*.

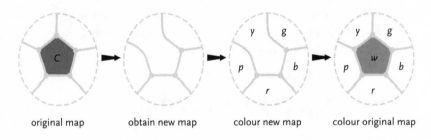

original map   obtain new map   colour new map   colour original map

We now reinstate the country *C*. Since six colours are available, and since the countries next to *C* use at most five of them, there must be a spare colour for colouring *C*. Thus, we can colour the entire map with six colours, which contradicts our assumption. This contradiction shows that no minimal criminal can exist, and proves the six-colour theorem.

. . . and Kempe solves it

We come now to the most famous fallacious proof in the whole of mathematics – the purported solution of the four-colour problem by the London barrister and amateur mathematician Alfred Bray Kempe (pronounced 'kemp'). It is unfortunate that he is now remembered mainly for his flawed proof, since he was a fine mathematician, highly regarded by his contemporaries. In Kempe's defence, his error was a subtle one that would remain undetected for eleven years, and his 'solution' embodies a number of original ideas that proved to be of the greatest importance in later work on the problem. Kempe's celebrated paper appeared in the *American Journal of Mathematics*, which is where we continue our story.

SYLVESTER'S NEW JOURNAL

Like his mathematical friend Arthur Cayley, with whom he worked so productively in the early 1850s, James Joseph Sylvester had great difficulty in securing an academic position. Until 1871, the professors at Oxford and Cambridge Universities were required to subscribe to the Thirty-nine Articles of the Church of England, and Sylvester, as a Jew, was not eligible for a position in either of the ancient universities. Indeed, though he achieved great success in

James Joseph Sylvester (1814–97)

his degree examinations while studying at St John's College, Cambridge, he was not allowed to receive the degree until many years later. From 1837 to 1841 he was Professor of Natural Philosophy at the non-sectarian University College, London, where he had briefly studied with Augustus De Morgan years earlier at the age of fourteen. It was not until 1855, after working for eleven years as an actuary at the Equity & Law Life Assurance Company in London, that he was able to secure a teaching post, at the Royal Military Academy at Woolwich. In 1870, new War Office regulations required all teaching staff at the military academies to retire on half-pay at the age of fifty-five, and Sylvester reluctantly left Woolwich to spend his remaining years in retirement – or so he thought.

For five years Sylvester spent his time publishing a book of poetry, singing in concerts (he took singing lessons from the French composer Charles Gounod) and dabbling in mathematics. But in late 1875 an event occurred that was to change his life. In America, the new Johns Hopkins University was being founded in Baltimore, Maryland, and its first president, Daniel Gilman, was determined to secure the best faculty members available. Sylvester, by this time possibly the finest mathematician in the English-speaking world, was approached and accepted this new challenge at a salary of five thousand dollars per annum, to be paid in gold.

Sylvester was blissfully happy at Johns Hopkins. He was able to develop his research ideas in a way that had been impossible at Woolwich, where routine teaching was the order of the day. In Baltimore he was surrounded by a group of idealistic and enthusiastic colleagues, such as the mathematician William Story and the astronomer Simon Newcomb, and he set himself the task of stimulating mathematical research activity at Johns Hopkins and, more widely, throughout the United States.

As part of his plans for this ambitious project, Sylvester founded the *American Journal of Mathematics* in 1878, with himself as editor-in-chief and Story as associate editor-in-charge. This publication, which still exists, was designed as a medium of communication between American mathematicians although 'its pages will always be open to contributions from abroad'. Indeed, Sylvester commissioned research articles from distinguished friends, both in America and overseas, and the first two volumes contained articles by Arthur Cayley and others from England, as well as papers from France, Germany and Denmark. Encouraged no doubt by Cayley, the editor-in-chief requested a paper from Alfred Kempe. The paper was entitled 'On the geographical problem of the four colours'.

KEMPE'S PAPER

Throughout his life, Alfred Bray Kempe was passionately interested in mathematics. At Cambridge he studied under Cayley at Trinity College, graduating in 1872, before embarking on a successful legal career. That same year he wrote his first mathematical paper, on the solution of equations by mechanical means, and five years later, stimulated by Peaucellier's discovery of a linkage for tracing a straight line (illustrated opposite), he published a popular memoir on linkages entitled 'How to draw a straight line'.

Kempe's work on linkages was so well regarded by his contemporaries that, when he was proposed for a Fellowship of the Royal Society, the first seven of the eight papers cited in support of his proposal were on this subject; the eighth was his paper on the four-colour problem. He later became the Royal Society's treasurer, a post he would hold for over twenty years, and he was knighted in 1912 for his many services to the Society. He was also a keen mountaineer, and Mount Kempe and the nearby Kempe Glacier in the Antarctic were named after him.

Kempe's interest in map colouring was aroused by Cayley's

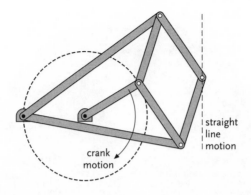

straight
line
motion

crank
motion

query at the meeting of the London Mathematical Society –
Kempe had attended the meeting – and by Cayley's subsequent
memoir in the Royal Geographical Society's *Proceedings* in April
1879. By June of that year, Kempe had obtained his solution of the
four-colour problem, and on 17 July he published a preview of it in
Nature. The full version appeared by the end of the year, in Volume 2 of the *American Journal of Mathematics*. On 26 February
1880 Kempe published simplified versions of his solution in
Nature and the *Proceedings of the London Mathematical Society*,
which corrected some minor errors in his original paper but left
intact the fatal flaw.

In his *American Journal of Mathematics* paper, Kempe started by
introducing the four-colour problem and remarking, as Cayley had
done, that De Morgan had 'stated somewhere' that the problem
was well known to mapmakers. He then mentioned Cayley's contributions, and continued:

Some inkling of the nature of the difficulty of the question, unless
its weak point be discovered and attacked, may be derived from the
fact that a very small alteration in one part of a map may render it
necessary to recolour it throughout. After a somewhat arduous
search, I have succeeded, suddenly, as may be expected, in hitting

Alfred Bray Kempe (1849–1922)

upon the weak point, which proved an easy one to attack . . . How this can be done I will endeavour – at the request of the Editor-in-Chief – to explain.

The rest of the paper falls into three main sections, followed by several shorter observations. In one section Kempe derived an extension of Euler's formula for maps, citing the corresponding plane version derived by Cauchy (see Chapter 3). He stated it thus:

in every map drawn on a simply connected surface the number of points of concourse [meeting points] and number of districts are together one greater than the number of boundaries

(that is, $V + F = E + 1$). From this, he deduced the formula

$$5d_1 + 4d_2 + 3d_3 + 2d_4 + d_5 - \ldots = 0,$$

where each d_k denotes the number of regions with k boundaries. This is similar to the counting formula we derived in Chapter 3, except that Kempe also allowed one-sided regions. Since the first five terms are the only positive ones, he deduced that not all the quantities d_1, d_2, d_3, d_4 or d_5 can be zero – in his words, 'every map drawn on a simply connected surface must have a district with less than six boundaries'. This is the result that we have called the 'only five neighbours' theorem.

Using this result, Kempe then described a method for colouring any map. It can be summarized in six steps:

1. Locate a country with five or fewer neighbours (such a country exists, by the above result).
2. Cover this country with a blank piece of paper (a *patch*) of the same shape but slightly larger.
3. Extend all the boundaries that meet this patch and join them together at a single point within the patch, as below, which amounts to shrinking the country to a point. It has the effect of reducing the number of countries by 1.

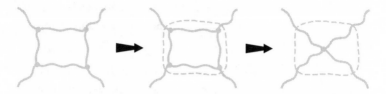

4. Repeat the above procedure with the new map, continuing until there is just one country left: the whole map is now said to be *patched out*.

5. Colour the single remaining country with any of the four colours.

6. Reverse the above process, stripping off the patches in reverse order, until the original map is restored. At each stage, colour the restored country with any available colour until the entire map is coloured with four colours.

It was in trying to carry out this final stage that Kempe made his most important contribution to map colouring. For a problem arises: how can we always be sure that one of the four colours is available when a country is restored? As we saw at the end of the previous chapter, there is no difficulty if our restored country has at most three boundary lines. For example, if it is a triangular country, then it is surrounded by three countries that are coloured with just three colours; there is then a fourth colour available to colour the triangle, as shown below. In the language of the previous chapter, this shows that no minimal criminal can contain a triangle.

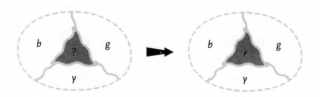

But what do we do if our restored country has four or five boundary lines – if it is a square or a pentagon? In either case, the restored country may be surrounded by countries that are coloured with all four colours, as illustrated below. There is then no colour available to colour the square or the pentagon.

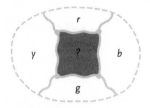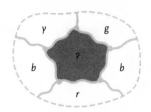

KEMPE CHAINS

To overcome this difficulty, Kempe introduced a method now known as the *method of Kempe chains* or a *Kempe-chain argument*. In this method, we look at the colours of the countries surrounding the central one and choose two of them that are not adjacent – say, *red* and *green*. We then look at only those countries that are coloured with these colours. We illustrate Kempe's method by first showing how he dealt with a square (which we denote by *S*), surrounded by four countries of different colours: this corresponds to the case where our minimal criminal contains a square.

We look first at the *red* and *green* neighbours of the square *S*. Each of these is the starting point for a *red–green* part of the map – that is, a part of the map consisting entirely of countries coloured *red* or *green*. (Although they are called *Kempe chains*, these two-coloured parts of the map are not usually chains, but may contain 'branches', as in the diagrams that follow: these branches may contain arbitrary arrangements of countries. As long as they are coloured legitimately, their presence has no bearing on the problem of how to colour *S*.)

We now ask whether these two *red–green* parts are separate from each other, or whether they link up. Two situations can arise:

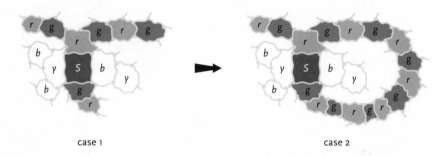

case 1   case 2

Case 1

Here the *red* and *green* countries above *S* that can be reached from the *red* neighbour of *S* do not link up with the *red* and *green* countries below *S* that can be reached from the *green* neighbour of *S*. We can therefore interchange the colours of the *red* and *green* countries above *S*, as shown below, without affecting the colouring of the *red* and *green* countries below *S*. The square *S* is then surrounded only by the colours *green*, *blue* and *yellow*, so that *S* can be coloured *red*. This completes the colouring of the map.

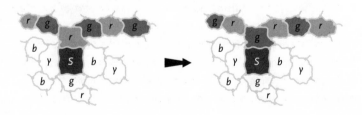

Case 2

Here the *red–green* part above *S* does link up with the *red–green* part below *S*. This makes things a little more difficult, since nothing is gained by interchanging the colours *red* and *green*: the *red* neighbour of *S* becomes *green*, and the *green* neighbour of *S* becomes *red*, and we are no better off than before.

We turn our attention instead to the *blue* and *yellow* countries, and to the *blue–yellow* parts of the map to the left and right of the square *S*. Here, the *blue–yellow* part to the right of *S* is cut off from the *blue–yellow* part to the left of *S*, because the chain of *red* and *green* countries gets in the way.

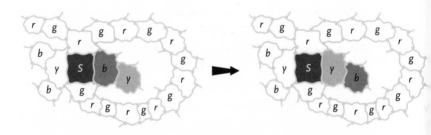

Thus, we can interchange the colours of the *blue* and *yellow* countries on the right of *S* without affecting the colouring of the *blue* and *yellow* countries on the left of *S*. The square *S* is then surrounded only by the colours *yellow*, *red* and *green*, so that *S* can be coloured *blue*.

This completes the colouring of the map when the restored country is a square, and shows that no minimal criminal can contain a square.

Kempe then turned his attention to the case where the restored country is a pentagon (which we denote by *P*), surrounded by five countries that are coloured with four different colours. (It is this part of the proof that contains the fundamental flaw, as we shall explain in Chapter 7.)

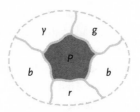

He again chose two of the surrounding colours that are not adjacent. He first considered the non-adjacent *yellow* and *red* countries above and below *P*. If the *yellow–red* part above *P* does not link up with the *yellow–red* part below *P*, then we can interchange the colours of the *yellow* and *red* countries above *P* without affecting the colouring of those below *P*.

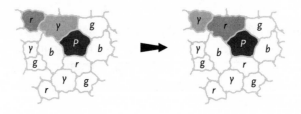

The pentagon *P* is then surrounded only by the colours *red*, *green* and *blue*, so that *P* can be coloured *yellow*, thereby completing the colouring of the map in this case. We are therefore left with the case where the *red–yellow* part above *P* links up with the *red–yellow* part below *P*.

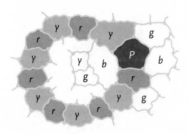

Kempe next considered the non-adjacent *green* and *red* countries above and below *P*. If the *green–red* part above *P* does not link up with the *green–red* part below *P*, then we can interchange the colours of the *green* and *red* countries above *P* without affecting the colouring of those below *P*.

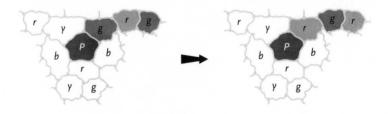

The pentagon *P* is then surrounded only by the colours *red*,
yellow and *blue*, so that *P* can be coloured *green*, completing the
colouring of the map in this case. We are therefore left with the
case where the *green–red* part above *P* links up with the *green–red*
part below *P*. Combining this with the previous case, we obtain:

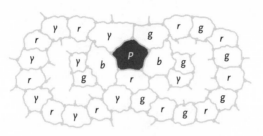

Notice that the *blue–yellow* part to the right of *P* is separated
from the *blue–yellow* part to the left of *P*, because the *red–
green* chain of countries gets in the way. We can thus interchange the colours of the *blue–yellow* part on the right of *P*
without affecting the colouring of the *blue–yellow* part on the
left of *P*.

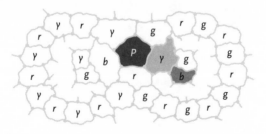

Similarly, the *blue–green* part to the left of *P* is separated from the *blue–green* part to the right of *P*, because the *red–yellow* chain of countries gets in the way. We can thus interchange the colours of the *blue–green* part on the left of *P* without affecting the colouring of the *blue–green* part on the right of *P*:

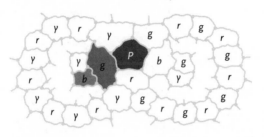

If we carry out both of these interchanges, then the pentagon *P* is surrounded only by the colours *yellow*, *red* and *green*, so *P* can be coloured *blue*.

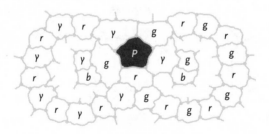

This completes the colouring of the map when the restored country is a pentagon, and shows that no minimal criminal can contain a pentagon.

Since we have now dealt with all possible cases, we have proved the four-colour theorem.

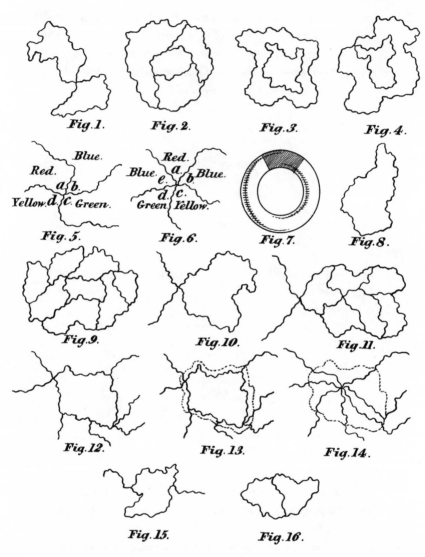

Fig.1.   Fig.2.   Fig.3.   Fig.4.

Blue.
Red.
Yellow. *a* *b* Green.
d *c*

Red.
Blue. *a* *b* Blue.
e
d *c*
Green. Yellow.

Fig.5.   Fig.6.   Fig.7.   Fig.8.

Fig.9.   Fig.10.   Fig.11.

Fig.12.   Fig.13.   Fig.14.

Fig.15.   Fig.16.

The diagrams from Kempe's paper in the American Journal of Mathematics.

SOME VARIATIONS

Although Kempe's proof was incorrect, as we shall explain in Chapter 7, his paper contained a number of further remarks. One of them concerns two special cases of interest that had not been mentioned by previous authors:

If every district of a cubic map is in contact with an even number of others along every circuit formed by its boundaries, three colours will suffice to colour the map.

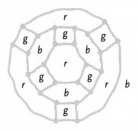

In this case, each country has an even number of neighbours – in particular, the map has no triangles or pentagons – so the countries surrounding any given country must alternate in colour.

If an even number of boundaries meet at every point of concourse [meeting point], two colours will suffice.

In this case, each meeting point is surrounded by an even number of countries, which must alternate in colour. Kempe made the further observation that such a map can be obtained by 'drawing any number of continuous lines crossing each other and themselves any number of times': for example, the map above is obtained by drawing three intersecting circles.

A second remark by Kempe is that the ideas of map colouring can be extended to maps drawn on surfaces other than the sphere. Like Charles Sanders Peirce in the 1860s, he remarked that maps drawn on the torus may require six colours – in fact, they may need up to seven colours, as we saw in Chapter 2. This idea of colouring maps on general surfaces would be considerably developed by Percy Heawood in 1890 (see Chapter 7).

Before leaving Kempe's classic paper, we should mention an important connection between map colouring and his work on linkages. As Kempe observed:

If we lay a sheet of tracing paper over a map and mark a point on it over each district and connect the points corresponding to districts which have a common boundary, we have on the tracing paper a diagram of a 'linkage', and we have the exact analogue of the question we have been considering, that of lettering the points in the linkage with as few letters as possible, so that no two directly connected points shall be lettered with the same letter.

This construction is similar to the one we used in Chapter 2, when we related the problem of the five palaces to Möbius's problem of the five princes. The linkage obtained from the map of mainland Australia is shown opposite. Any colouring of the countries of the map gives rise to a lettering of the points in the linkage in which no two directly connected points are lettered the same.

We now refer to such a linkage as a *graph* (a term coined in 1878 by Sylvester in a chemical context, and having no connection with the usual use of that term) and to the above process as forming

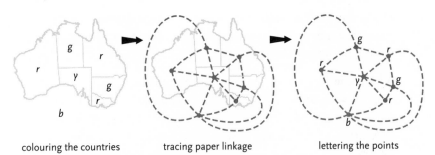

colouring the countries   tracing paper linkage   lettering the points

the graph (or dual graph) of the map. This reformulation of the four-colour problem as a problem involving the lettering of points reappeared briefly in the 1880s (see Chapter 6) and was later reintroduced in the 1930s and used in all subsequent attempts to solve the problem.

So as not to complicate matters, we shall stick to colouring the countries of maps (rather than switching to lettering the points of a graph, now or at a later stage), throughout the rest of this book.

BACK TO BALTIMORE

In the meantime, back at Johns Hopkins University, Kempe's paper had provoked much interest. On 5 November 1879, William Story presented the main ideas of Kempe's proof to an audience of eighteen at a meeting of the Johns Hopkins Scientific Association. Story suggested a few small corrections to Kempe's work, extending the section on Euler's formula to some types of map that Kempe had overlooked. Furthermore, in order 'to make the proof absolutely rigorous', he appended a four-page 'Note on the preceding paper' to Kempe's paper in the *American Journal of Mathematics*. It is a pity that Story's improvements did not extend to pointing out the fundamental error.

Story's 'Note' brought him into dispute with the testy Sylvester who, considering its publication inappropriate and believing his associate editor to have acted unprofessionally, wrote an angry letter to President Gilman:

When I consider Story's conduct since my absence this year and couple it with the fact of his disobeying my directions when I was absent last year and an inexcusable want of right feeling not to say *mala fides* exhibited by him in his treatment of Mr. Kempe's valuable memoir, I have come to the conclusion that it is inexpedient that we should continue to act together in carrying on the *Journal* . . .

Fortunately, Gilman was able to calm things down, but Story's name ceased to appear on the *Journal*'s title page as 'Associate Editor in Charge'.

Also present at the November gathering was Charles Sanders Peirce, by now a part-time lecturer in logic at Johns Hopkins University, while also working for the U.S. Coastal Survey. Prompted no doubt by Kempe's manuscript, Peirce had written to Story on 17 August mentioning that he had just come across his old solution of the four-colour problem – possibly the purported proof that he had presented at Harvard University in the 1860s (see Chapter 2). At the 5 November meeting at which Kempe's proof was discussed, 'Remarks were made upon this paper by Mr. C. S. Peirce', and at the following meeting on 3 December, Peirce

discussed a new point in respect to the Geographical Problem of the Four Colours, showing by logical argumentation that a better demonstration of the problem than the one offered by Mr. Kempe is possible.

The details of Peirce's 'better demonstration' have not survived, but his recasting of the four-colour problem in algebraic terms is outlined below. Here, A, B, C and D represent colours, and 1, 2 and 3 are countries. A symbol such as c_2 means that colour C is assigned to country 2, and he wrote the 'equation' $c_2 = 1$ if country 2 is so coloured and $c_2 = 0$ if it is not. Peirce then reformulated the four-colour problem as follows (the symbol i refers to any of the countries 1, 2 or 3). An explanation is given immediately after it.

$A_i{}^2 - A_i = 0$; and similarly for B, C, D;

$A_iB_i = 0$; and similarly for all pairs made up of A, B, C, D.

Also $A_i + B_i + C_i + D_i = 1$.

And if i and j are the numbers of two regions having a common boundary, $A_iA_j = 0$, as does the product of any similar pair of B's, C's, D's.

In the first line, the equation $A_i{}^2 - A_i = 0$ can be solved to give $A_i = 1$ or 0 – that is, colour A is either assigned to country i, or it is not – and similarly for each of the other colours B, C and D. The second line asserts that colours A and B cannot both be assigned to country i, and similarly for any other pair of colours. The third line means that one of the colours A, B, C or D must be assigned to country i. The fourth line tells us that if the countries i and j have a common boundary line, then colour A cannot be assigned to both – and similarly for colours B, C and D.

Further publicity for Kempe's proof appeared in the pages of the Christmas Day 1879 issue of the New York periodical *The Nation*. Included among its book reviews and descriptions of concerts, plays and operas was a discussion of the current *American Journal of Mathematics*. This featured an explanation of the 'curved ball in baseball' and a 'Notice by Mr. Peirce' concerning the first proof of 'a mathematical proposition known to be true many years before any one succeeded in producing a demonstration of it', by 'Mr. Kempe who was well-known for his investigation into linkage'.

Peirce and Story continued to be interested in the four-colour problem for the rest of their lives. Peirce lectured on it at a meeting of the National Academy of Sciences in New York on 15 November 1899, and there are many jottings of maps and their

colourings among his various notebooks at Harvard University.
We conclude this chapter with an extract from a letter from Story
to Peirce that well illustrates the frustrations that the problem
caused them.

Dec. 1, 1900

My dear Peirce:

. . . As to my not answering your letter about the four-color problem, I am heartily tired of that subject. I have spent an immense
amount of time on it, and all to no purpose. Your first method had
occurred to me years ago, but I did not succeed in getting anything
out of it.

Dec. 6, 1900

My delay in sending this off is largely your own fault. You have
again reminded me of that fascinating but elusive problem, and I
have spent the time since writing the above in trying to solve it, but
alas! I believe that the case of exception to Kempe's method
requires that the map shall have at least one triangular or quadri-
lateral district, in which case the pentagon is not the next district to
be colored, i.e. the exception does not occur. But I cannot prove
it . . .

   Sincerely Yours,
   William E. Story

A chapter of accidents

Kempe's proof of the four-colour theorem was widely accepted, and quickly became part of mathematical folklore. We have seen how it led to discussions at Johns Hopkins University in Baltimore, and it seems to have been universally accepted by British mathematicians also, including Arthur Cayley and, as we shall see, Peter Guthrie Tait. Kempe was duly proposed as a Fellow of the Royal Society on 24 November 1879 and was elected on 2 June 1881.

One Victorian Englishman who enjoyed the four-colour problem was Lewis Carroll, author of *Alice's Adventures in Wonderland* and *Through the Looking Glass*. Under his real name of Charles Lutwidge Dodgson, he lectured in mathematics for many years at Christ Church, Oxford. A keen advocate of traditional approaches to geometry, and of the study of Euclid's *Elements* in particular, he also carried out pioneering work in an area of mathematics known as symbolic logic. The story is told (which Dodgson always denied) that Queen Victoria was so enchanted by the *Alice* books that she asked for his next book to be sent to her. When she was handed Dodgson's *An Elementary Treatise on Determinants*, she was not amused.

Dodgson enjoyed inventing puzzles and games and showing them to the many children of his acquaintance, such as Alice

Liddell and her sisters, for whom he wrote the *Alice* books. In particular, according to his nephew Stuart Collingwood, a favourite puzzle of Dodgson's was the following:

A is to draw a fictitious map divided into counties.

B is to colour it (or rather mark the counties with names of colours) using as few colours as possible.

Two adjacent counties must have different colours.

A's object is to force B to use as many colours as possible.

How many can he force B to use?

In France, a translation of Kempe's paper by Edouard Lucas appeared in the journal *Revue scientifique* in 1883. Eleven years later, shortly after Lucas's death, an extended version of it appeared in the final volume of his celebrated four-volume work, *Récréations mathématiques*.

In Germany, at Richard Baltzer's 1885 Leipzig Scientific Society lecture on Möbius's problem of the five princes (see Chapter 2), the distinguished German mathematician Felix Klein drew Baltzer's attention to 'the relevant work of Mr. Kempe in London, on the geographical problem of the four colours'. As we have seen, Baltzer confused the two problems, claiming that his proof of the impossibility of five neighbouring regions provided a much simpler method for solving the four-colour problem, and concluding, 'How delighted Möbius would have been to see such a far-reaching application of his friend Weiske's economical proof.' (Weiske was the colleague who apparently suggested to Möbius the problem of the five princes.) Unfortunately, as we now see, Baltzer was not the only person who was accident-prone.

A CHALLENGE FOR THE BISHOP

On 1 January 1887, the following paragraphs appeared in the *Journal of Education*. They were sent in by the Revd James Maurice Wilson, a pioneer in the teaching of science in schools and the Headmaster of Clifton College, a boys' school in the Bristol area.

One of the 'institutions' at Clifton College is, that the Headmaster sets a Challenge Problem to the whole school once a term. Sometimes these problems refer to mechanical inventions, or applications of electricity; sometimes mathematical problems, which require ingenuity rather than knowledge for their solution. One of his mathematical problems, for example, was to prove that, if all the angular points of a regular decagon are joined, and all the sides and diagonals produced indefinitely, the number of triangles so formed will be 10,000.

He has been good enough to send us the Challenge Problem for last term, which we append. Perhaps it will tempt some of our correspondents to send us a solution.

'In colouring a plane map of counties, it is of course desired that no two counties which have a common boundary should be coloured alike; and it is found, on trial, that four colours are always sufficient, whatever the shape or number of the counties or areas may be. Required, a good proof of this. Why four? Would it be true if the areas are drawn so as to cover a whole sphere?

'Solutions to be sent to the Head Master on or before Dec. 1 . . . No solution may exceed one page, 30 lines of MS., and one page of diagrams.'

The Clifton College challenge problem caused much interest and was taken up in exalted circles. In the 1 June 1889 issue of the *Journal of Education*, Wilson wrote again, saying:

Some time ago there appeared in your columns a statement of a
problem which attracted a good deal of interest . . . Many attempts
have been made, and some mathematicians have written to me to
say that they could not satisfy themselves with any proof which
they had arrived at; and I have been frequently asked whether any
neat and satisfactory solutions have been given. I think, therefore,
that many of your readers will be glad to see the solution which I
now send. It is due to the Bishop of London. He writes:– 'I wrote
the enclosed at an evening meeting of the ———— ————, while a
vehement ———— was saying what I did not care to listen to.'

The next morning's papers reported that:

The Bishop of London was greatly interested in Mr. ————'s
speech, and was observed to be taking notes all the time he was
speaking.

Frederick Temple, Bishop of London and later Archbishop of
Canterbury, had once taught mathematics at Balliol College,
Oxford, and was particularly interested in mathematical games
and puzzles. Unfortunately, his 'solution', which was given in full,
shows that, like Richard Baltzer and so many others, Temple had
considered it sufficient to prove that it is impossible to draw five
mutually neighbouring regions in the plane. Temple's misunder-
standing was explained in detail several years later by John Cook
Wilson, Wykeham Professor of Logic at Oxford University.

A VISIT TO SCOTLAND

Another person who was accident-prone was Peter Guthrie Tait, professor of natural philosophy in Edinburgh and co-author, with William Thomson (later Lord Kelvin), of a widely read *Treatise on Natural Philosophy*. Tait was a distinguished mathematical physicist, as well as a passionate golfer, and his interest in the models of trajectories and the behaviour of materials under impact led to a classic paper on the trajectory of a golf ball. After carrying out a number of practical experiments, he managed to calculate the greatest theoretical distance that a golf ball can travel, and presented his results to the Royal Society of Edinburgh. According to an anecdote related in *Golf* magazine, his son Frederick Guthrie Tait, the finest amateur golfer of his time, then proceeded to hit his golf ball five yards further than the greatest distance predicted by his father! However, this report is more likely to have been the product of a journalist's fertile imagination.

Tait was particularly interested in Kempe's note in *Nature* of 26 February 1880 on the solution of the four-colour problem. Tait had first learned of the problem from Cayley some years earlier, while working on some mathematical problems involving knots, and had independently obtained Kempe's result that if there are an even number of boundary lines at each meeting point in a map, then two colours suffice (though, as he observed, 'boundaries usually meet in threes').

Tait considered Kempe's solution of the four-colour problem to be unnecessarily lengthy, giving 'little insight into its real nature and bearings', and in less than three weeks he produced four or five simpler solutions of his own – none of them correct. These he confidently presented to the Royal Society of Edinburgh on 15 March, publishing his results in a paper in its *Proceedings*.

Tait's first abortive approach to the problem used Kempe's

Peter Guthrie Tait (1831–1901)

linkage diagrams, with points representing countries, and lines connecting points corresponding to neighbouring countries. He added further lines to divide the diagram into triangles – for, if the points of the 'triangulated diagram' can be labelled with four letters so that no two connected points are lettered the same, then (by removing the added lines) so can the points of the original diagram.

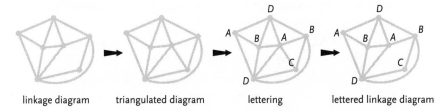

linkage diagram   triangulated diagram   lettering   lettered linkage diagram

Tait now added extra points so as to make all the triangular countries four-sided; all the points can then be alternately lettered A and B. Then, Tait said, 'do the same thing again *another way*', giving a second lettering. Finally, he superimposed these two lettered maps, omitting all added points.

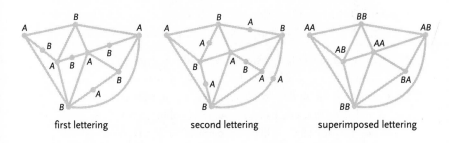

first lettering   second lettering   superimposed lettering

Tait claimed that, for every map, this last diagram gives a lettering of the points in the four composite 'letters' AA, AB, BA and BB, in which any two connected points are lettered differently. However, it is by no means certain whether the lettering can always be carried out. Tait admitted as much in a letter to Kempe on 13 April 1880, written hastily while he was invigilating an examination:

There is at first a little puzzle about the two separate ways of lettering the diagram — so as to make the two copies different all over — but it is not difficult to get over that by the help of a few simple rules.

Since he did not explain his 'few simple rules', or indeed why his method works, his attempt is not a satisfactory solution of the four-colour problem.

Later in 1880, Tait presented his one really useful and original idea on the subject, which he believed would yield a solution of the four-colour problem. It did not, but it led to an interesting area of study that is still active today. Starting with a cubic map, he considered colouring not its countries but its boundary lines:

In a map where only three boundaries meet at each point, the boundaries may be coloured with three colours, so that no two of the same colour are coterminous.

(Two boundary lines are *coterminous* if they share a common endpoint.) For example, consider any cubic map whose countries are coloured A, B, C and D. We can then colour the boundary lines with the three colours *a* (alpha), *β* (beta) and *γ* (gamma), as follows:

use colour *a* for all lines between pairs of countries coloured
   A and B, or between pairs of countries coloured C and D;
use colour *β* for all lines between pairs of countries coloured
   A and C, or between pairs of countries coloured B and D;
use colour *γ* for all lines between pairs of countries coloured
   A and D, or between pairs of countries coloured B and C.

The procedure is illustrated below:

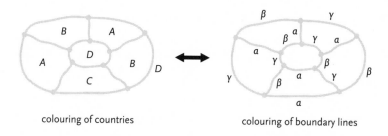

colouring of countries   colouring of boundary lines

This process can always be reversed. Suppose that we are given a cubic map in which the boundary lines have been coloured in such a way that all three colours appear at each meeting point. Choose any country and colour it with colour A. The above recipe then tells us the colours of the immediately neighbouring countries: for example, if the boundary line between the original country and a neighbouring one is coloured β, then the neighbouring country must be coloured c. Continuing in this way, we can colour all the countries of the map.

An alternative way of reversing the process, which will be of importance later, is to choose just two of the colours and look at the boundary lines with these colours: they always form one or more 'closed cycles'.

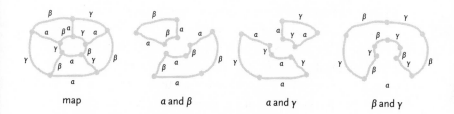

map   α and β   α and γ   β and γ

We now take the first of these sets of cycles and write 1 for all countries inside the cycles, and 0 for all countries outside. We repeat this for the second set of cycles, and superimpose the results:

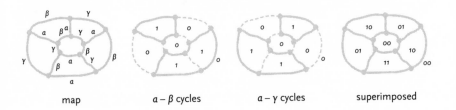

map   α – β cycles   α – γ cycles   superimposed

This always produces a colouring of the countries in the four 'colours' 00, 01, 10 and 11.

Tait correctly realized the importance of colouring the boundary lines of a map, and wrote a further letter to Kempe. He considered his earlier result, that the boundary lines of a cubic map can always be coloured with three colours, to be

really the proper thing to start with; as a Lemma easily proved.

So I have withdrawn my long paper and read a much shorter & simpler one (based on the above Lemma) at the last meeting of the R. S. E.

I think I now see the simple principle on wh. the whole depends.

Tait presented this result to the Royal Society of Edinburgh, on 19 July 1880, and an abstract duly appeared in the Society's *Proceedings*. Tait believed that his 'Lemma easily proved' could be proved easily by mathematical induction (see Chapter 4), and that a solution of the four-colour problem could then be deduced from it, as above. Unfortunately, his 'Lemma easily proved' is just as difficult to prove as the original four-colour theorem.

There is a postscript to Tait's early writings on the four-colour problem. His first paper inspired another 'Note on the colouring of maps' in the Royal Society of Edinburgh's *Proceedings*. It began:

From the *Proceedings of the Royal Society of Edinburgh*, No. 106, p. 501, it appears the colouring of maps is receiving attention. This note bears chiefly upon the history of the matter.

The writer was Frederick Guthrie, and this was the note in which he identified his brother Francis as the originator of the four-colour problem, as we saw in Chapter 2.

CYCLING AROUND POLYHEDRA

Tait's abstract in the Royal Society of Edinburgh's *Proceedings* was soon expanded into a paper that appeared in the Society's *Transactions*. In this 'Note on a theorem in geometry of position', he returned to colouring the boundary lines of a cubic map:

If $2n$ points be joined by $3n$ lines, so that three lines, and three only, meet at each point, these lines can be divided (usually in many different ways) into three groups of n each, such that one of each group ends at each of the points.

For example, in the following coloured diagram with 8 points and 12 lines (corresponding to $n = 4$), the lines coloured a form one of the three groups, the lines coloured β form the second, and the lines coloured γ form the third.

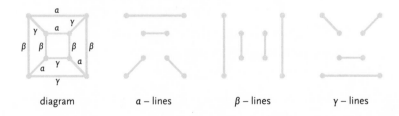

diagram a – lines β – lines γ – lines

Tait then went on to make the confusing assertion that:

The difficulty of obtaining a simple proof of this theorem originates in the fact that it is not true without limitation.

His example of where the result fails was the following diagram with 14 points and 21 lines (corresponding to *n* = 7). However, this diagram cannot correspond to a map, because the line in the middle does not separate two *different* countries:

Tait excluded such unwelcome situations by concentrating on cubic maps that can be obtained by projecting a polyhedron onto a plane. For a polyhedron to project to a cubic map it must be a *cubic polyhedron*, such as the tetrahedron, cube or dodecahedron, in which exactly three edges (or faces) meet at each vertex; the octahedron and icosahedron are not cubic polyhedra, since more than three edges (or faces) appear at each vertex. Tait believed his result to be 'universally true' for all cubic polyhedra, asserting that

the four-colour theorem implies, and is implied by, the statement that the boundary lines of these polyhedral maps can be coloured with three colours so that all three colours appear at each intersection point.

As a way of attacking this 'restricted problem', Tait asked the following question:

for each cubic polyhedron, is there always a single closed cycle that passes through all the vertices just once?

For the dodecahedron and the truncated octahedron, the following closed cycles pass through all the vertices just once:

dodechahedron   truncated octahedron

Tait's interest in these cycles arose from the fact that if a cubic polyhedron has such a cycle, then we can colour its edges with three colours by alternating the colours *red* and *green* along the cycle and colouring the remaining edges *blue*. For example, we obtain the following colouring of the edges of a cube:

cube   cycle   colouring of edges

In fact, this question of finding a cycle that passes through all the vertices of a polyhedron had been studied over twenty years earlier by the Reverend Thomas Penyngton Kirkman and Sir William Rowan Hamilton. We now digress briefly and look at their contributions.

A VOYAGE AROUND THE WORLD

Thomas Kirkman was rector of the tiny parish of Croft-with-Southworth, near Warrington in Lancashire. His parochial duties were not demanding, and left him time to have seven children and do a lot of mathematics, becoming a Fellow of the Royal Society in the process. Kirkman was fascinated by polyhedra, and tried to

Thomas Penyngton Kirkman
(1806–95)

find a closed cycle that passes just once through each of the vertices of a given polyhedron. Unfortunately, his papers are not easy to follow since he invented his own terminology, such as a *p-edral q-acron* for a polyhedron with *p* faces and *q* vertices, and a *triedral summit* (in his spelling) for a vertex at which three faces meet.

In 1855, Kirkman found a polyhedron that has no such cycle. In his words:

If we cut in two the cell of a bee by a section of its six parallel edges, we have a 13-acron, whose faces are one hexagon and nine quadrilaterals. The closed 13-gon cannot be drawn.

This polyhedron can be projected onto the plane as follows:

To see why there is no cycle passing through all thirteen vertices, we colour the vertices *black* and *white* in such a way that each edge has a *black* end and a *white* end. Any cycle then has to alternate between *black* vertices and *white* ones, and this is possible only if the numbers of *black* and *white* vertices are the same. But there are seven *black* vertices and six *white* ones, so no cycle is possible. However, as Kirkman observed, if we join the central vertex to the one on its left, then we can draw the required cycle.

The following year, William Rowan Hamilton (whom we met briefly in Chapter 2) became interested in drawing cycles on a dodecahedron as a consequence of his work on quaternions – his so-called *icosian calculus*. In particular, he considered three quantities, represented by the Greek symbols ι (iota), κ (kappa) and λ (lambda), that satisfy the equations

$\iota^2 = 1$, $\kappa^3 = 1$ and $\lambda^5 = 1$, where $\lambda = \iota\kappa$.

Writing μ (mu) $= \iota\kappa^2$, Hamilton was able to show that $\mu^5 = 1$. He also obtained the lengthy expression

$\lambda^3\mu^3\lambda\mu\lambda\mu\lambda^3\mu^3\lambda\mu\lambda\mu = 1$, which he then proceeded to interpret in terms of cycles on a dodecahedron:

if we start at the vertex B, proceed towards C, and at each junction think of λ as 'turn right' and μ as 'turn left', then we obtain the required cycle with the letters in alphabetical order:

B C D F G H J K L M N P Q R S T V W X Z,

ending back at B.

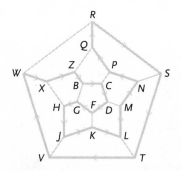

Hamilton was so proud of his icosian calculus that he converted it into *A Voyage Around the World*, a 'new and highly amusing game for the drawing room' in which the vertices are labelled with the twenty consonants B, C, D, . . . , X, Z, representing the places Brussels, Canton, Delhi, . . . , Xerez, Zanzibar. The object is to 'travel around the world', visiting these cities along the way, and return to one's starting point; one solution is to proceed in alphabetical order, as indicated above. Hamilton 'found that some young persons have been much amused' by trying his game, and proudly sold it for £25 to the games manufacturer John Jaques and Son of Hatton Garden, London. In the event, this was a wise move on Hamilton's part, as it did not sell.

The Icosian Game had twenty numbered pegs that were to be placed into the holes B, C, D, . . . in cyclic order. The instructions

THE ICOSIAN GAME.

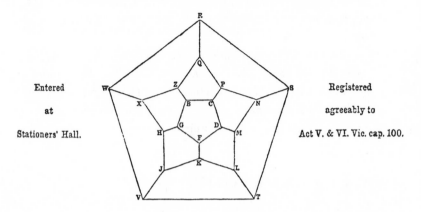

Entered

at

Stationers' Hall.

Registered

agreeably to

Act V. & VI. Vic. cap. 100.

In this new Game (invented by Sir WILLIAM ROWAN HAMILTON, LL.D., &c., of Dublin, and by him named *Icosian*, from a Greek word signifying "twenty") a player is to place the whole or part of a set of twenty numbered pieces or men upon the points or in the holes of a board, represented by the diagram above drawn in such a manner as always to proceed *along the lines* of the figure, and also to fulfil certain *other* conditions, which may in various ways be assigned by another player. Ingenuity and skill may thus be exercised in *proposing* as well as in *resolving* problems of the game. For example, the first of the two players may place the first five pieces in any five consecutive holes, and then require the second player to place the remaining fifteen men consecutively in such a manner that the succession may be *cyclical*, that is, so that No. 20 may be adjacent to No. 1; and it is always possible to answer any question of this kind. Thus, if B C D F G be the five given initial points, it is allowed to complete the succession by following the alphabetical order of the twenty consonants, as suggested by the diagram itself; but after placing the piece No. 6 in the hole H, as before, it is *also* allowed (by the supposed conditions) to put No. 7 in X instead of J, and then to conclude with the succession, W R S T V J K L M N P Q Z. Other Examples of Icosian Problems, with solutions of some of them, will be found in the following page.

LONDON:

PUBLISHED AND SOLD WHOLESALE BY JOHN JAQUES AND SON, 102 HATTON GARDEN;

AND TO BE HAD AT MOST OF THE LEADING FANCY REPOSITORIES

THROUGHOUT THE KINGDOM.

The original instructions for Hamilton's Icosian Game.

for the game included various puzzles of the form, 'given five initial points, in how many ways can the cycle be completed?' For example, how many cycles start with *BCDFG* (there are two), or with *LTSRQ*, or with *JVTSR*?

Such were Sir William's importance and influence that these cycles are now called *Hamiltonian cycles*, rather than being more justly credited to Kirkman, who had beaten Hamilton to them by several months and had also investigated cycles on general polyhedra – not just on a dodecahedron.

WEE PLANETOIDS

We now return to the work of Peter Guthrie Tait. As we saw earlier, Tait was interested in Hamiltonian cycles on cubic polyhedra because they enabled him to colour the edges with three colours, and thereby to colour the faces with four colours. In his November 1880 paper, Tait discussed his 'theorem' that every cubic poly-hedron has such a cycle, and observed that:

HAMILTON's *Icosian Game* is a particular application of this theorem, the corresponding figure being a projection of a pentag-onal dodecahedron. It was suggested to him by the remark, in Mr KIRKMAN's paper on Polyhedra (*Phil. Trans.* 1858, p. 160), that a clear 'circle of edges' of a unique type passed through all the sum-mits of this polyhedron.

Tait was incorrect in saying that Kirkman gave Hamilton the idea for the Icosian Game.

Three years later, Tait followed this paper with a lecture to the Edinburgh Mathematical Society and a further paper, in the *Philo-sophical Magazine*, in which he reaffirmed his view that all cubic polyhedra have Hamiltonian cycles, remarking:

Probably the proof of this curious proposition has hitherto escaped detection from its sheer simplicity. Habitual stargazers are apt to miss the beauties of the more humble terrestrial objects.

Kirkman had meanwhile returned to the subject of cycles on polyhedra in 1881, perhaps encouraged by Tait's interest in his work. He recast the problem of Hamiltonian cycles on cubic polyhedra in verse form and sent it to the *Mathematical Questions and Solutions of the Educational Times*. Victorian mathematicians frequently expressed their problems in verse, and Kirkman was no exception. Written in the form of a truly appalling piece of doggerel of 53 lines, *Question 6610* began:

6610. (By the Rev. T. P. KIRKMAN, M.A., F.R.S.) –
To a wee planetoid, but recently out,
I am bound to attach an opinion
On how to effect a design they're about
Of improving their little dominion.
Tired of their islands, they long for a Continent:
Here is the statement they give me, their want anent –

and concluded:

On islands, three, four, or two,
Towns, to threescore or two,
Cover with triedral summits your n-edron;
When they are penned, run
Over your islands a pencilling cloud,
Giving the cities the shore-lines allowed.
The white will all be
Ferry, cable, and sea.
You may feel a bit proud,
If, after some labour, you find what you want, an ent-
Ire single circle of towns on a Continent.

When it is found, there is nothing to face,
But proof of a rule to fit every case.

The 'Solution by the PROPOSER' commences with the words:

The theorem – that on every *p*-edron *P*, having only triedral summits, a closed circle of edges passes once through every summit, has this provoking interest, that it mocks alike at doubt and proof.

In other words, it is extremely difficult to decide whether every cubic polyhedron has a Hamiltonian cycle: Kirkman's cell-of-a-bee polyhedron has no Hamiltonian cycle, but it does not constitute a counter-example because it is not a cubic polyhedron.

Kirkman went on to give a plausible proof of the statement, concluding that:

This is the only attempt to prove the general theorem that I desire now to offer. Demonstration it is not; but it will probably carry conviction until a *p*-edron is produced, on which the sought circle cannot be found. A rigorous demonstration is much to be desired, flowing from the simple definition of a *p*-edron with triedral summits. But I share the opinion of Professor P. G. TAIT, that our prospect of obtaining such demonstration is very remote indeed. It is my impression that he knows more about these circles than any other.

In the event, it would be a further sixty-five years before the *impossibility* of finding such a proof was demonstrated. In 1946, the cubic polyhedron opposite was discovered by the English mathematician Bill Tutte. It contains no Hamiltonian cycle, thereby proving the unfortunate Tait wrong yet again.

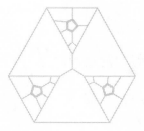

We conclude this chapter by remarking that, like Kempe, Tutte's name is pronounced with a silent 'e', inspiring a colleague to pen the following words:

> Beware the displeasure of Tutte:
> He is normally equable, but
> He gets in a temper
> When people say 'Kemp-e',
> And mutters 'Not Tutt-e, but Tutte.'

7 A bombshell from Durham

Kempe's proof of the four-colour theorem was a very good one. It was incorrect, but it was a very good incorrect proof. Not only did it convince Victorian mathematicians for eleven years, but most of the ideas it contained were sound and would form the basis for later assaults on the problem, including the one that was ultimately to prove successful. The credit for finding the error in Kempe's paper goes to Percy Heawood (pronounced 'haywood'), a mathematics lecturer at the Durham Colleges (now Durham University).

HEAWOOD'S MAP

Percy John Heawood studied at Exeter College, Oxford, gaining degrees in both mathematics and classics and winning the University's junior and senior mathematics prizes. In 1887 he was appointed 'mathematical lecturer' at Durham, where he spent the rest of his life, becoming professor and eventually vice-chancellor, and retiring at the age of seventy-eight in 1939.

Throughout his life, Heawood had a passion for committees and considered a day as wasted if he did not attend at least one committee meeting. A noted Latin, Greek and Hebrew scholar, he

Percy John Heawood (1861–1955)

continued with his studies in mathematics and the classics after his retirement, and in his ninetieth year published two research papers – one on the four-colour problem in the *Proceedings of the London Mathematical Society*, the other on the date of the Last Supper in the *Jewish Quarterly Review*.

Like several of the other characters in our story, Heawood was somewhat eccentric. As his London Mathematical Society obituary notice affectionately observed:

In his appearance, manners and habits of thought, Heawood was an extravagantly unusual man. He had an immense moustache and a meagre, slightly stooping figure. He usually wore an Inverness cape of strange pattern and manifest antiquity, and carried an ancient handbag. His walk was delicate and hasty, and he was often accompanied by a dog, which was admitted to his lectures.

One particularly unusual trait of 'Pussy' Heawood (so nicknamed because his moustache resembled a cat's whiskers) was that he used to set his watch just once a year, on Christmas Day. Whenever he needed to know the time, he would do the necessary calculations in his head, based on how quickly it lost time. He reportedly once confided to a colleague who had asked him the time, 'No, it's not two hours fast; it's ten hours slow!'

In addition to his work on map colouring, which is described below, Heawood's major achievement was to save the magnificent eleventh-century Durham Castle, set on a promontory above the River Wear, from sliding down the cliff into the water below. In 1928 its foundations were found to be insecure, and money was hard to find. Almost single-handedly, as secretary of the Castle Preservation Committee, Heawood managed to raise the necessary funding and the castle was saved. For these efforts Durham University awarded him an honorary Doctorate of Civil Laws in 1931, and he was created an Officer of the Order of the British Empire (OBE) in 1939.

Percy Heawood's interest in the four-colour problem dated back to his first term at Oxford University. In a letter to the Belgian mathematician Alfred Errera in the 1920s, he recalled that:

When I went up as a student to Oxford in 1880, H. J. S. Smith was the Professor of Geometry. A very clear and interesting lecturer, before proceeding with the proper subject material of his course he started by dividing geometrical properties into (1) properties of situation, (2) descriptive properties, (3) metrical properties; and among two or three examples of (1), he cited the four colour theorem as probably true but unproved; since then, this problem has always fascinated me.

We saw in Chapter 4 that Henry Smith had presided over the London Mathematical Society meeting at which Cayley raised his query about map colouring, but Smith was evidently unaware of Kempe's subsequent solution. Heawood's letter continued:

I thus turned naturally, when I heard about it, to the so-called proof of Kempe – and after a critical examination, I discovered the error I indicated in my printed note. I had never heard of any doubts raised as to its validity before my paper in the 'Quarterly Journal'.

Heawood wrote his famous paper 'Map-colour theorem' in June 1889, and it duly appeared in the *Quarterly Journal of Mathematics* in June of the following year. In his introduction, Heawood seemed almost apologetic about uncovering the error in Kempe's paper:

The present article does not profess to give a proof of this original Theorem; in fact its aims are so far rather destructive than constructive, for it will be shown that there is a defect in the now apparently recognized proof . . .

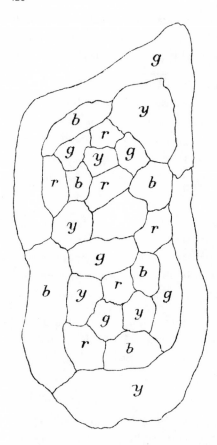

Heawood's counter-example to Kempe's proof.

Heawood's demolition of Kempe's proof centres on the argument that Kempe used when dealing with a pentagon. As we saw in Chapter 5, Kempe considered an arrangement of countries like this:

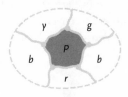

(Here the colours surrounding the pentagon have been changed so as to agree with the example in this chapter.) Kempe then used two simultaneous interchanges of colour to recolour the countries on either side of the pentagon so that the pentagon itself could then be coloured. Either interchange of colour is perfectly valid, but to do both at once is not permissible.

To explain why this is so, Heawood introduced the map opposite. In this map, each of the twenty-five countries has been coloured *red*, *blue*, *yellow* or *green*, except for the pentagon (which we shall call *P*) in the middle. This map can certainly be coloured with four colours, but the point of Heawood's example is to show that Kempe's *method of proof* is incorrect.

Following Kempe, we now try to recolour two of the pentagon's neighbours in such a way that there is a spare colour available for *P*. We notice first that the *blue* and *yellow* neighbours of *P* are connected by a *blue–yellow* chain of countries that separates the *red–green* part above *P* from the *red–green* part below *P*, as shown in figure (a) overleaf. We can thus interchange the colours of the *red–green* part above *P* without affecting the *red–green* part below *P*, as in figure (b).

Alternatively, we could have carried out a different interchange of colours. The *blue* and *green* neighbours of *P* are connected by a *blue–green* chain of countries that separates the *red–yellow* part above *P* from the *red–yellow* part below *P*, as shown in figure (c) overleaf. We can thus interchange the colours of the *red–yellow* part below *P* without affecting the *red–yellow* part above *P*, as in figure (d).

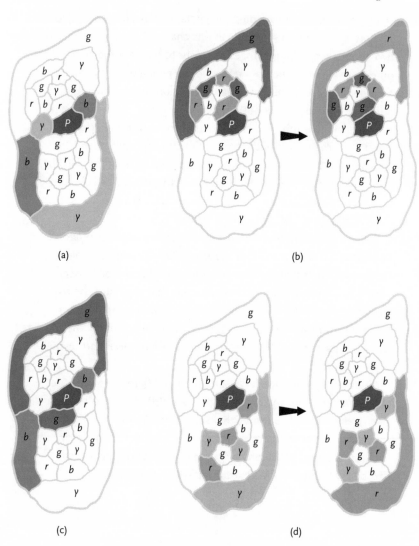

(a)

(b)

(c)

(d)

Either of these colour interchanges is permissible on its own, but Kempe's error was in trying to do them simultaneously. For, if we interchange the colours in both the *red–green* part above *P* and the *red–yellow* part below *P*, then the *green* country *A* and the *yellow* country *B* both become *red*, which is not permissible.

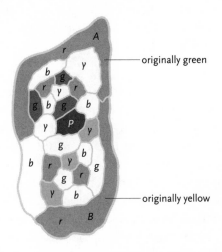

Thus Kempe's proof, which allowed the possibility of such simultaneous colour interchanges, is fallacious. In Heawood's own words:

But, unfortunately, it is conceivable that though either transposition would remove a red, both may not remove both reds. My map is an actual exemplification of this possibility, where either transposition prevents the other from being of any avail, by bringing the red and the other divisions into the same region; so that Mr. Kempe's proof does not hold, unless some modifications can be introduced into it to meet this case of failure.

The flaw in Kempe's argument proved to be a major one. Alfred Kempe admitted his error in the columns of the *Proceedings of the*

London Mathematical Society, and on 9 April 1891 he discussed the
difficulty at a meeting of the Society:

My proof consisted of a method by which any map could be
coloured with four colours. Mr. Heawood gives a case in which the
method fails, and thus shows the proof to be erroneous. I have not
succeeded in remedying the defect, though it can be shown that the
map which Mr. Heawood gives can be coloured with four colours,
and thus his criticism applied to my proof only and not to the
theorem itself.

In fact, Heawood's map is not the simplest that he could have
given, as was shown by two Belgian mathematicians. A more
attractive example has the form of the buckyball C_{30} (see Chapter
3), and was described by Alfred Errera in 1921. A version of it
appears below; as before, we can carry out either of the *red–green*
and *red–yellow* colour interchanges, but not both. Earlier, in 1896,
Charles-Jean-Gustave-Nicolas de la Vallée Poussin, known particularly for his work on the distribution of prime numbers and his
proof of the prime number theorem, had presented an example.

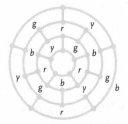

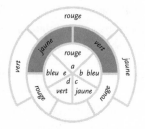

Errera's example   de la Vallée Poussin's example

He had read Edouard Lucas's translation of Kempe's proof in *Récréations mathématiques* and, unaware of Heawood's paper, had
located Kempe's error. Here, the pentagon appears as a single
point and we can interchange *either* the countries coloured *bleu*

and *jaune* on the left, *or* those coloured *bleu* and *vert* on the right, but if we do both, then the two shaded countries both become *bleu*.

A SALVAGE OPERATION

Although Heawood was unable to mend the hole in Kempe's proof, he salvaged enough from Kempe's ideas to prove the *five-colour theorem*. Although weaker than the four-colour theorem, it is still a remarkable result:

Five-colour theorem

Every map can be coloured with at most five colours in such a way that neighbouring countries are coloured differently.

To prove the five-colour theorem, we imitate Cayley and Kempe's approach to the four-colour problem in Chapters 4 and 5. That is, we begin by assuming that the five-colour theorem is false, so that there are some maps that *cannot* be coloured with five colours – and among these special maps that need five colours or more, we consider a minimal criminal – one with the smallest possible number of countries. This map *cannot* be coloured with five colours, but any map with fewer countries *can* be coloured with five colours.

We now use the 'only five neighbours' theorem from Chapter 3. This theorem tells us that our map must contain a country with at most five neighbours – a digon, triangle, square or pentagon. If there is a digon, triangle or square, the proof is simple. We illustrate the method of proof for a map that contains a square. (The proofs for a digon or a triangle are similar but easier, and are

Four Colours Suffice

almost identical to the proof of the four-colour theorem for a map
containing a digon or a triangle given in Chapter 4.)

Suppose that our minimal criminal contains a square, as shown
below. If we remove a boundary line from the square, merging the
square with one of its former neighbours, we obtain a map with
fewer countries. By our assumption, we can colour this new map
with five colours, *red*, *blue*, *green*, *yellow* and *orange*.

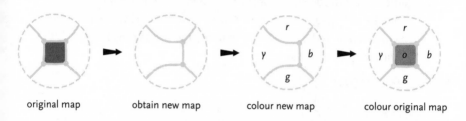

original map   obtain new map   colour new map   colour original map

We now reinstate the square. Since five colours are available,
and since the countries next to the square use only four of them,
there must be a spare colour for the square. Thus, we can colour
the minimal criminal with five colours, which contradicts our
assumption. This shows that a minimal criminal cannot contain a
square.

If our minimal criminal contains a pentagon, then, as before, we
remove a boundary line and merge the pentagon with its former
neighbour, giving us a map having fewer countries. By our
assumption, we can colour this new map with five colours
(below).

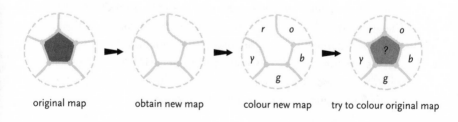

original map   obtain new map   colour new map   try to colour original map

As before, we next reinstate the pentagon. But now the countries next to the pentagon may use all five colours, so there is no spare colour to colour the pentagon.

To rescue the situation we use a Kempe-chain argument (see Chapter 5). We choose two of the surrounding colours that are not adjacent – say, *red* and *green* – and look at only those countries that are coloured with these colours. (From here on, the proof is almost identical to the proof of the four-colour theorem for a map containing a square.)

We look first at the *red* and *green* neighbours of the pentagon (which we call *P*). Each of these is the starting point for a part of the map consisting entirely of countries coloured *red* or *green*. Now, are these two *red–green* parts separate from each other, or do they link up?

Two cases can arise:

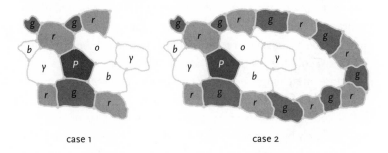

case 1   case 2

Case 1

In the first case, the *red* and *green* countries above *P* that can be reached from the *red* neighbour of *P* do not link up with the *red* and *green* countries below *P* that can be reached from the *green* neighbour of *P*. We can therefore interchange the colours of the *red* and *green* countries above *P*, as shown below, without affecting the colouring of the *red* and *green* countries below *P*. The pentagon *P* is then surrounded only by the colours *green*, *blue*, *yellow* and *orange*, so *P* can be coloured *red*. This completes the colouring of the map.

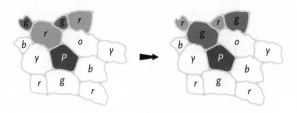

Case 2

In this case, where the *red–green* part above *P* does link up with the *red–green* part below *P*, nothing is gained by interchanging the colours. So we turn our attention to the *blue* and *yellow* countries, and to the *blue–yellow* parts of the map to the left and right of the pentagon *P*. Here, the *blue–yellow* part to the right of *P* is cut off from the *blue–yellow* part to the left of *P*, because the chain of *red* and *green* countries gets in the way, as illustrated below. We can therefore interchange the colours of the *blue* and *yellow* countries on the right of *P* without affecting the colouring of the *blue* and *yellow* countries on the left of *P*. The pentagon *P* is then surrounded only by the colours *yellow*, *red*, *green* and *orange*, so *P* can be coloured *blue*. This completes the colouring in the second case.

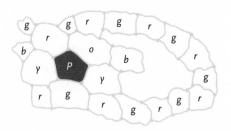

Thus, in either case, the minimal criminal can be coloured with
five colours, which contradicts our assumption. This shows that
no minimal criminal can contain a pentagon, and completes the
proof of the five-colour theorem.

COLOURING EMPIRES

Percy Heawood was keen to extend the idea of map colouring
beyond the original four-colour problem. In the introduction to his
paper in the *Quarterly Journal of Mathematics*, he pointed out that
his main object was neither to call attention to the error in
Kempe's proof, nor to prove the five-colour theorem: it was rather
to contrast with the four-colour problem 'some remarkable
generalizations of it, of which strangely the rigorous proof is much
easier'. One of these was the *empire problem*. It can be thought of
as the problem of colouring a map with several 'empires': each
empire consists of a 'mother country' and a number of 'colonies'
that must all be coloured the same as the mother country.

In Chapter 1 we gave the following example of a map, due to
Heawood, in which one country splits into two pieces and more
than four colours are needed.

He now asked how many colours are needed if 'any country may consist of two distinct portions but no more', so that each empire consists of a mother country and possibly a single colony. His simple example used five colours, but it is possible to construct maps that need many more colours than this. Using an argument based on Euler's formula, he showed that the number of colours required does not exceed twelve. But are there maps that actually do need all twelve colours?

Heawood presented the example below, 'obtained with much difficulty in a more or less empirical manner'. It consists of twelve pairs of countries, with the property that each pair has a single boundary line in common with a country from every other pair. For example, of the two countries numbered 8, one has boundary lines in common with countries numbered 1, 2, 6, 7, 9, 10 and 12, while the other has boundary lines in common with countries numbered with the remaining numbers, 3, 4, 5 and 11. Thus, there are twelve 'mutually neighbouring pairs of countries', which therefore need twelve colours.

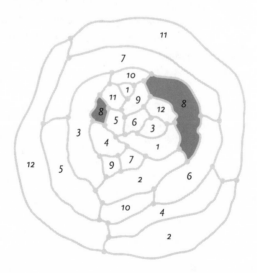

Again, using an argument based on Euler's formula, Heawood proved that if each empire consists of three distinct portions, then at most eighteen colours are needed; and, more generally, for any number *r* greater than 1, if each empire consists of *r* distinct portions, then at most 6*r* colours are needed. However, apart from the case *r* = 2 illustrated above, Heawood was unable to find maps for which these bounds are attained.

It was more than ninety years before any progress was made in solving the empire problem. In 1981, the case *r* = 3 was solved by Herbert Taylor, who constructed the eighteen-colour map below. The solution for a general value of *r* was obtained three years later, by Brad Jackson and Gerhard Ringel.

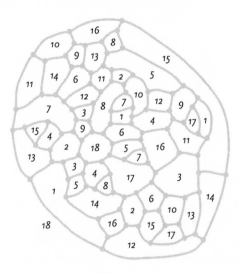

MAPS ON DOUGHNUTS

Heawood also investigated the colouring of maps drawn on
surfaces other than the sphere. In his 1890 paper he presented an
example of a map drawn on the surface of a torus (or doughnut)
in which each country adjoins every other country. Such a torus
map, similar to Heawood's, is shown below – we came across it
in Chapter 2, in connection with Möbius's problem of the
princes.

This map has seven mutually neighbouring countries, and colouring it therefore requires seven colours. But can all maps on the
torus be coloured with seven colours?

To answer this, we need a version of Euler's formula that
applies to maps drawn on the torus. We recall from Chapter 3
that, for maps drawn on the plane or the surface of a sphere,
Euler's formula takes the following form:

(number of countries) − (number of boundary lines)
+ (number of meeting points) = 2

or, in symbols,

$F - E + V = 2.$

For maps drawn on the surface of a torus, Euler's formula is similar, except that 0 appears on the right-hand side:

Euler's formula for the torus
(number of countries)
− (number of boundary lines)
+ (number of meeting points) = 0
or, in symbols,
$F - E + V = 0.$

This result is essentially the same as Lhuilier's formula for a polyhedron with one hole tunnelled through it (see Chapter 3). As an example, the above map on the torus has 7 countries, 21 boundary lines and 14 meeting points, so $F = 7$, $E = 21$, $V = 14$, and Euler's formula becomes $F - E + V = 7 - 21 + 14 = 0$.

Just as every map on the plane or sphere has at least one country with five or fewer neighbours, so we have the following result for a torus:

'Only six neighbours' theorem for the torus
Every map on a torus has at least one country with six or fewer neighbours.

To prove this, we imitate the proof we gave in Chapter 3, as follows.

Consider a map on the torus with F countries, E boundary lines and V meeting points. As we did before, we can assume that there are at least three boundary lines at each meeting point, and we again obtain the inequality $V \leq \frac{2}{3}E$.

To prove that the map has at least one country with six or fewer neighbours, we assume the opposite – that our map has no such country. It follows from this assumption that each country is surrounded by at least seven neighbours. So, counting up the boundary lines around all the countries, we appear to get at least $7F$ lines because each of the F countries is surrounded by at least seven lines. But each line has been counted twice, because it has a country on each side, so we must divide by 2. So E is at least $\frac{7}{2}F$: in symbols, $E \geq \frac{7}{2}F$, which we can rewrite as $F \leq \frac{2}{7}E$.

If we now put these inequalities $F \leq \frac{2}{7}E$ and $V \leq \frac{2}{3}E$ into Euler's formula for a torus, we obtain

$$F - E + V \leq \frac{2}{7}E - E + \frac{2}{3}E = -\frac{1}{21}E.$$

But by Euler's formula, $F - E + V$ is equal to 0. It follows that $0 \leq -\frac{1}{21}E$, which is clearly false. Our error has arisen from our assumption that each country has at least seven neighbours – so this assumption is unjustified. It follows that at least one country has six or fewer neighbours.

Using the 'only six neighbours' theorem, we can now prove our main result – that every map on the torus can be coloured with seven colours. To do this, we assume that we have a minimal criminal: such a map *cannot* be coloured with seven colours, but any map with fewer countries *can* be coloured with seven colours. By the above result, this map has a country C with six or fewer neighbours. If we now remove a boundary line from the country C, merging C with one of its neighbours, then we obtain a map with fewer countries. By our assumption, we can colour this new map with seven colours, which we choose to be *red, blue, green, yellow, orange, white* and *purple*:

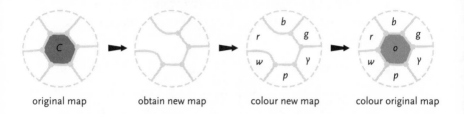

original map   obtain new map   colour new map   colour original map

We now reinstate the country C by replacing the removed line. Since seven colours are available, and since the countries next to C use at most six of them, there must be a spare colour that can be used to colour C. We can therefore colour the minimal criminal with seven colours, which contradicts our assumption. This shows that every map on a torus can be coloured with seven colours.

We can summarize the above discussion thus:

every map on a torus can be coloured with seven colours – and there are torus maps that need seven colours.

Heawood proved these results in his paper, and then tried to extend them to doughnuts with more than one hole (or 'pretzels'). For example, how many colours are needed to colour maps on the following two-holed torus?

By arguments similar to those used above, we can prove that the 'magic number' is 8:

> every map on a two-holed torus can be coloured with eight colours – and there are two-holed torus maps that need eight colours.

We can continue in this way:

> every map on a three-holed torus can be coloured with nine colours – and there are three-holed torus maps that need nine colours;
>
>   . . .
>
> every map on a ten-holed torus can be coloured with fourteen colours – and there are ten-holed torus maps that need fourteen colours;

and so on.

For maps on a torus with an arbitrary number of holes, Heawood used Lhuilier's version of Euler's formula:

Euler's formula for the *h*-holed torus

(number of countries)

− (number of boundary lines)

+ (number of meeting points) = 2 − 2*h*

or, in symbols,

$$F - E + V = 2 - 2h.$$

This gives rise to a rather complicated-looking result:

> every map on a torus with *h* holes can be coloured with $H(h)$ colours, where $H(h)$ is the number
>
> $[\frac{1}{2}(7 + \sqrt{1 + 48h})].$

Here, the square brackets mean that we round the number down if it is not already a whole number – for example,

$[7] = 7$ and $[9.99] = 9$.

So, when $h = 1$ (a torus), the number of colours is

$H(1) = [\frac{1}{2}(7 + \sqrt{49})] = [7] = 7;$

when $h = 2$ (a two-holed torus), the number of colours is

$H(2) = [\frac{1}{2}(7 + \sqrt{97})] = [8.42 \ldots] = 8;$

and when $h = 10$ (a ten-holed torus), the number of colours is

$H(10) = [\frac{1}{2}(7 + \sqrt{481})] = [14.46 \ldots] = 14.$

These values agree with the numbers given above.

The number $H(h)$ is sometimes called the *Heawood number* of the h-holed torus. A table of values of $H(h)$ for small values of h is as follows:

number of holes, h	1	2	3	4	5	6	7	8	9	10
number of colours, $H(h)$	7	8	9	10	11	12	12	13	13	14

Unfortunately, even Heawood was capable of making mistakes. As we have seen, he correctly proved the formula for maps on the torus, showing that seven colours are sufficient, and he gave an example of a map on a torus that needs seven colours. For larger values of h, he indeed showed that every map on an h-holed torus can be coloured with $[\frac{1}{2}(7 + \sqrt{1 + 48h})]$ colours. He then asserted that there are maps that require this number of colours – but he neglected to prove this, commenting merely that

for highly connected surfaces it will be observed that there are generally contacts enough and to spare for the above number of divisions each to touch each.

Heawood's omission was a major blunder. It was first noticed in 1891, the year after Heawood published his paper, by Lothar Heffter of Giessen in Germany. Heffter managed to show that there are indeed maps that require $H(h) = [\frac{1}{2}(7 + \sqrt{1 + 48h})]$ colours when $h = 2, 3, 4, 5, 6$, and a few other values. But he was unable to prove the result in general. As we shall see in Chapter 9, seventy-seven years would pass before what came to be known as the *Heawood conjecture* was finally proved:

Heawood conjecture

For each positive number h, there is a map on the surface of an h-holed torus that requires $H(h) = [\frac{1}{2}(7 + \sqrt{1 + 48h})]$ colours.

Although neither Heawood nor Heffter lived to see the proof of the conjecture, their doughnut colouring activities seem to have bestowed longevity: Heawood died at the age of ninety-four, while Heffter survived to the ripe old age of ninety-nine.

PICKING UP THE PIECES

Heawood's paper seems to have gone largely unnoticed. In the final volume of Edouard Lucas's *Récréations mathématiques*, published posthumously in 1894, an extended version of Kempe's paper appeared, but with no mention of his error. In *L'Intermédiaire des mathématiciens*, a newly formed Paris publication for the dissemination of mathematical problems, P. Mansion posed the four-colour problem, seemingly unaware of any of the previous work on it. Replies were soon forthcoming – by H. Delannoy and A. S. Ramsey (citing papers by Kempe and Tait), by de la Vallée

Poussin (presenting his simpler example of a map that exposes Kempe's error) and again by Delannoy (who did not understand de la Vallée Poussin's example and thought that Kempe had been right all along!).

The *Intermédiaire* discussion then moved on to Tait's assertion that all cubic polyhedra have Hamiltonian cycles – cycles passing through all the vertices. In 1898–9, the Danish mathematician Julius Petersen wrote two short notes, commenting on the connections between Tait's work and the four-colour theorem and remarking that 'M. Kempe only skimmed over the problem; he committed his error just where the difficulties began.' Petersen concluded with a surprising opinion: 'I know nothing with certainty, but if it came to a wager I would maintain that the theorem of the four colours is not correct.'

Petersen is now mainly remembered for the *Petersen graph*, a diagram that appears below in its usual form, in Petersen's own 1898 drawing from the *Intermédiaire*, and in a version given twelve years earlier by Kempe. It can be shown that the Petersen graph has no Hamiltonian cycle, and that it does not arise from a polyhedron.

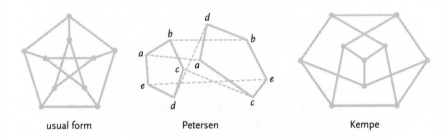

usual form   Petersen   Kempe

At no stage in the *Intermédiaire* correspondence were Heawood's achievements mentioned. Undeterred, Heawood continued to struggle with the four-colour problem for the next sixty years. In 1898 he wrote a paper that included a particularly useful

development of Tait's connection (see Chapter 6) between colouring the boundary lines of a cubic map with three colours and
colouring its countries with four colours:

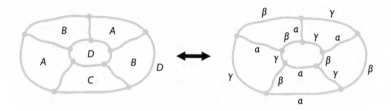

Heawood now focused on the meeting points, to which he
assigned the value 1 if the colours a, β and γ occur in clockwise
order, and −1 if they occur in anticlockwise order. For example, the
top point in the diagram above is assigned the value 1, because
the colours a, β and γ appear there in clockwise order. This gives
the following arrangement of 1's and −1's.

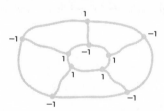

Heawood showed that the sum of the numbers around each
country of the map must always be divisible by 3: here, the sum is
3 for the country in the middle, 0 for each country in the ring, and
−3 for the exterior country. Turning the argument around, he then
showed that

if we can assign the numbers 1 and −1 to the meeting points of a
cubic map so that the sum of the numbers around each country is

divisible by 3, then we can colour the boundary lines with three colours, and the countries with four colours.

Thus, solving the four-colour problem amounts to showing that we can always assign the numbers 1 and −1 to the meeting points in this way. Heawood was so fascinated by this approach that he wrote five further papers on it, but without achieving the ultimate success that he so much craved.

There is a simple consequence of Heawood's result, which Heawood himself seems to have been the first to notice. As he observed in his 1898 paper: 'If the number of neighbours of each country is divisible by 3, then the map can be coloured with four colours.' To see why, assign the number 1 to every meeting point. Then, by the condition on the neighbours, the sum of the numbers around each country must be divisible by 3, and so the colours a, β and γ must appear in clockwise order around each point. We can then obtain the required colouring of the countries, as explained in Chapter 6.

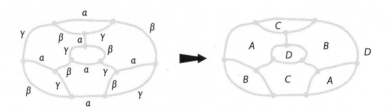

We conclude this chapter with a further result of Heawood's. In a paper of 1936, written when he was in his mid-seventies, he attempted to estimate the probability that the four-colour problem was true − or, more precisely, the probability that a random map with n countries could be coloured with four colours. His arguments were somewhat crude, but they indicated that the probability of failure did not exceed about $e^{-4n/3}$, where e is the exponential number 2.71828 . . . and n is fairly large. By 1936 the four-colour theorem was known to be true for all maps with $n = 27$ countries, as we shall see in Chapter 8, so his arguments would imply that the likelihood of failure was less than one in a million billion. It followed that if the four-colour theorem were false, then a minimal criminal would be very hard to find.

143

Crossing the Atlantic

The end of the nineteenth century marks a watershed in the fortunes of the four-colour problem. Kempe's solution had been shown to be faulty, and nothing had turned up to take its place. It was time for some new ideas to burst upon the scene.

The feeling was also beginning to emerge in some quarters that a solution to the four-colour problem had not been forthcoming because no really good mathematicians had worked on it. Indeed, a story is told about the distinguished German number-theorist Hermann Minkowski in the first decade of the twentieth century. While lecturing on topology at Göttingen University, he mentioned the four-colour problem:

'This theorem has not yet been proved, but that is because only mathematicians of the third rank have occupied themselves with it', Minkowski announced to the class in a rare burst of arrogance. 'I believe I can prove it.'

He began to work out his demonstration on the spot. By the end of the hour he had not finished. The project was carried over to the next meeting of the class. Several weeks passed in this way. Finally, one rainy morning, Minkowski entered the lecture hall, followed by a crash of thunder. At the rostrum, he turned towards the class, a deeply serious expression on his face.

'Heaven is angered by my arrogance', he announced. 'My proof of the Four Colour Theorem is also defective.' He then took up the lecture on topology at the point where he had dropped it several weeks before.

With the turn of the century, things began to change. The four-colour saga ceased to be a largely British story as several American mathematicians – George Birkhoff, Oswald Veblen, Philip Franklin, Hassler Whitney, and others – contributed to a major new chapter. Through their work, two ideas gradually emerged that were to prove decisive. Both ideas had been implicit in Kempe's paper – the concepts of an *unavoidable set* and a *reducible configuration*.

TWO FUNDAMENTAL IDEAS

In Chapter 3 we proved the 'only five neighbours' theorem – that every map has at least one country with five or fewer neighbours. It follows, in particular, that every cubic map must contain at least one of the following:

digon

triangle

square

pentagon

So whenever we draw a cubic map, we have to use at least one country from this collection. We describe such a collection of countries as an *unavoidable set*, because we cannot avoid it – in any cubic map at least one of them must appear somewhere. As we shall see later, another unavoidable set is this one:

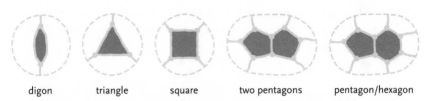

digon   triangle   square   two pentagons   pentagon/hexagon

Thus, if a cubic map contains no digon, triangle or square, then not only must it contain a pentagon, it must contain either two adjacent pentagons or a pentagon adjacent to a hexagon. So why should we be interested in unavoidable sets?

In trying to prove the four-colour theorem, our approach has been to consider a minimal criminal – a cubic map with the smallest possible number of countries that *cannot* be coloured with four colours – and to prove that minimal criminals cannot exist. In Chapter 4 we showed that no minimal criminal can contain a digon or triangle, since if it did, any colouring of the rest of the map with four colours could immediately be extended to include the digon or triangle as well. In Chapter 5 we gave Kempe's proof that no minimal criminal can contain a square – if there were such a country, then, by using a Kempe-chain argument that interchanges two of the colours in part of the map, we could free up a colour for the square. But Kempe was unable to prove satisfactorily that no minimal criminal can contain a pentagon: the error in his proof was pointed out by Heawood, as we saw in Chapter 7.

So the smallest country there can be in a minimal criminal is a pentagon. Moreover, as we saw in our discussion of the counting formula in Chapter 3, the map must contain at least twelve pentagons. It follows that a minimal criminal must have at least twelve countries. However, if a cubic map has *exactly* twelve countries, then they must all be pentagons, and we obtain the map of the dodecahedron. But this map *can* be coloured with four colours, as shown below, so it cannot be a minimal criminal. It follows that every map with up to twelve countries can be coloured with four colours, and hence that a minimal criminal must have at least thirteen countries.

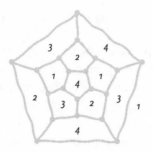

A *reducible configuration* is any arrangement of countries that cannot occur in a minimal criminal – so a digon, a triangle and a square are all reducible configurations. If a map contains a reducible configuration, then any colouring of the rest of the map with four colours can be extended, after any necessary recolouring, to a colouring of the entire map. If Kempe had also been able to prove that the pentagon is reducible, then the four-colour problem would have been solved.

The rest of this book is concerned with the attempt to find *an unavoidable set of reducible configurations*. Finding such a set proves the four-colour theorem: since the set is unavoidable, every map must contain at least one of the configurations, but each configuration is reducible and so cannot be contained in a minimal criminal. Thus, no minimal criminal can exist. But how do we find unavoidable sets, and how do we find reducible configurations?

FINDING UNAVOIDABLE SETS

We have just seen how Kempe failed to deal with the pentagon. Can we replace the pentagon by some other arrangements of countries that we can handle more successfully? The first attempts to do this were made by Paul Wernicke, a German mathematician who received a doctoral degree from Göttingen University and then crossed the Atlantic to take up a professorship at the University of Kentucky. At one stage he was in charge of the University's military training, holding a commission as a colonel in the Kentucky militia; it is not recorded whether, in common with another Kentucky colonel, he had a liking for fried chicken.

In August 1897 Wernicke presented a paper entitled 'On the solution of the map-color problem' at the Fourth Summer Meeting of the American Mathematical Society, held in Canada at the University of Toronto. The Society's *Bulletin* included a short abstract of his paper, from which it appears that he was concerned with developing Tait's connection (described in Chapter 6) between colouring the boundary lines of a cubic map with three colours and colouring its countries with four colours. Wernicke's idea seems to have been to add new countries to a map so as to turn it into one he could manage:

Given a map correctly colored and with its frontiers marked, the author proves that any triangles, quadrangles, and pentagons can be introduced and correctly marked at the same time. The main theorem then follows by induction.

However, it is unlikely that Wernicke's proof of the four-colour theorem, by the method of induction, was any more successful than Tait's had been seventeen years earlier.

Wernicke's next attempts were much more productive. In a lengthy paper written in Göttingen in May 1903, he proved that a

cubic map that contains no digon, triangle or square must contain
either two adjacent pentagons or a pentagon adjacent to a hexagon. Thus, as we saw above, the following configurations of countries form an unavoidable set.

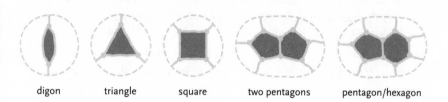

digon   triangle   square   two pentagons   pentagon/hexagon

To explain why Wernicke's result is true, we use a modern
approach based on the so-called *method of discharging*. This
method was originated by Heinrich Heesch (pronounced 'haish')
and first published in 1969; Heesch's work takes centre stage in
Chapter 9. The term *discharging* was introduced by Wolfgang
Haken, whose contributions to the eventual solution of the four-
colour problem are described in Chapter 10.

We illustrate the method of discharging by showing that the
above set of configurations – the digon, triangle, square, pair of
adjacent pentagons, and pentagon adjacent to a hexagon – form
an unavoidable set. To do this, we first assume that we have a
cubic map that contains none of them, and then seek a contradiction. The assumption tells us that no pentagon can be adjacent to
a digon, a triangle or a square (because there are none), or to
another pentagon or a hexagon. Thus, each pentagon can adjoin
only countries bounded by at least seven edges.

We now assign to each country a number – which we can think
of as an electrical charge. This we do as follows: to each country
with k boundary lines, we assign a charge of $6 - k$, so that

each pentagon ($k = 5$) receives a charge of 1;
each hexagon ($k = 6$) receives zero charge;

each heptagon ($k = 7$) receives a charge of -1;
each octagon ($k = 8$) receives a charge of -2;

and so on.

So if the map has c_5 pentagons, c_6 hexagons, c_7 heptagons, and so on, then the total charge on the map is

$$(1 \times c_5) + (0 \times c_6) + (-1 \times c_7) + (-2 \times c_8) + (-3 \times c_9) + \ldots$$
$$= c_5 - c_7 - 2c_8 - 3c_9 - \ldots \tag{1}$$

We now recall the counting formula from Chapter 3: if a cubic map has c_2 digons, c_3 triangles, c_4 squares, and so on, then

$$4c_2 + 3c_3 + 2c_4 + c_5 - c_7 - 2c_8 - 3c_9 - \ldots = 12.$$

Since our map has no digons, triangles or squares, we have $c_2 = c_3 = c_4 = 0$, and the counting formula simplifies to

$$c_5 - c_7 - 2c_8 - 3c_9 - \ldots = 12. \tag{2}$$

Comparing equations (1) and (2), we find that the total charge on the map is 12, a positive number.

We now move the charges around the map, in such a way that no charge is created or destroyed – this is called *discharging the map*, and is akin to the conservation of electrical charge. One way of doing this is to transfer one-fifth of a unit of charge from each pentagon to each of its five negatively charged neighbours – those with seven or more sides. The diagram below shows an example of this:

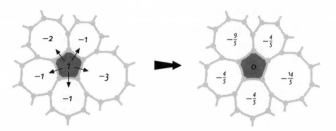

The result of the discharging is that the total charge on the map remains 12, but each pentagon now has zero charge, and each hexagon still has zero charge.

What happens to a heptagon? For a heptagon with initial charge -1 to receive enough contributions of $\frac{1}{5}$ to acquire positive charge, it would need at least six neighbouring pentagons, as shown below, supplying $\frac{6}{5}$ in total – but then at least two of these pentagons would have to be adjacent, which is not allowed. After the discharging, then, each heptagon retains its negative charge.

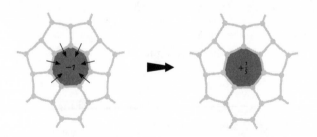

And what happens to an octagon? For an octagon with initial charge -2 to receive enough contributions of $\frac{1}{5}$ to acquire positive charge, it would need at least eleven neighbouring pentagons, which is clearly impossible. So, after the discharging, each octagon retains its negative charge – and so does each nonagon, decagon, and so on.

Thus, after the discharging, each country of our map has zero or negative charge. But this is inconsistent with the map having a total charge of 12. This contradiction proves that every cubic map must contain at least one of the above configurations of countries – a digon, a triangle, a square, two adjacent pentagons, and a pentagon adjoining a hexagon – and so these five configurations form an unavoidable set.

The idea of an unavoidable set was later developed by two other mathematicians. The first was Philip Franklin, who wrote a doctoral thesis on map colouring at Princeton University, and who later became a distinguished mathematician at the Massachusetts Institute of Technology (MIT). He was a brother-in-law of Norbert Wiener, the inventor of cybernetics – the comparative study of communication processes and automatic control systems.

In 1920 Franklin presented part of his thesis to the National Academy of Sciences. His presentation included the result, which he derived from the counting formula, that every cubic map must contain at least one digon, triangle, square, or one of the following:

a pentagon adjacent to two other pentagons;
a pentagon adjacent to a pentagon and a hexagon;
a pentagon adjacent to two hexagons.

This gives rise to the unavoidable set with nine configurations shown overleaf.

The other provider of unavoidable sets was the French mathematician Henri Lebesgue, better known for developing the *Lebesgue integral* used in mathematical analysis. In 1940, the year before his death, Lebesgue wrote a paper on some simple consequences of Euler's formula in which he used the counting formula to construct a number of interesting new unavoidable sets.

By modifying the method of discharging we can show that many

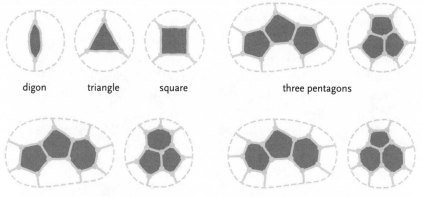

| digon | triangle | square | three pentagons |

two pentagons and a hexagon   a pentagon and two hexagons

sets of configurations are unavoidable, but the details of the discharging process may vary from one situation to another. Earlier we transferred one-fifth of a unit of charge from each pentagon to each negatively charged neighbour, but in other cases it may be more advantageous to transfer one-fourth or one-third of a unit, or to distribute the charge on each pentagon equally among *all* its neighbours.

As the twentieth century progressed, unavoidable sets with thousands of configurations were constructed. To deal with such enormous sets it proved necessary to keep on modifying the discharging process until it could deal with every possible case that arose. In Chapter 10 we shall see how this was done.

FINDING REDUCIBLE
CONFIGURATIONS

We have seen above that the digon, triangle and square are all
examples of reducible configurations – if they appear in a map,
then any colouring of the rest of the map with four colours can be
extended, possibly after some recolouring, to include them. But so
far our supply of reducible configurations has been very restricted.
This situation changed dramatically in 1913 with the publication of
a paper by George David Birkhoff.

Birkhoff, one of the outstanding American scholars of the early
twentieth century, made substantial contributions to many areas
of mathematics. He studied at the University of Chicago and at
Harvard University, later obtaining positions at the University of
Wisconsin and at Princeton University before returning to Harvard
in 1912 to spend the rest of his productive life.

While at Princeton, Birkhoff attended seminars by the celebrated geometer Oswald Veblen, who took a keen interest in the
four-colour problem. On 27 April 1912 Veblen read a paper to the
American Mathematical Society, extending the ideas of Percy
Heawood's 1898 paper and placing them in the context of a
particular type of geometry in which each line contains exactly
four points. Veblen was later to supervise Philip Franklin's doctoral thesis on the colouring of maps.

From this time onward, Birkhoff regarded solving the four-
colour problem as one of his greatest aspirations, even though he
was later to regret the amount of time he had spent on it. In 1913
he published his ground-breaking paper in the *American Journal
of Mathematics*, the journal in which Kempe's famous solution
had appeared over thirty years before. This paper, 'The reducibility
of maps', written while Birkhoff was still at Princeton, gave a

George David Birkhoff (1884–1944)

systematic treatment of Kempe-chain arguments and laid the
groundwork for all later developments in this direction.

Birkhoff considered *rings of countries* in minimal criminals. To
see what is involved, suppose that a map contains a ring of three
countries, with at least one other country inside the ring and at
least one other country outside it (the ring of countries is shaded):

Since the entire map is a minimal criminal, we can colour the
ring and its interior with four colours, and we can also colour the
ring and its exterior. We can then match up the three colours on
the ring (by renaming the colours, if necessary) and thereby obtain
a colouring of the whole map (see overleaf).

It follows that rings of three countries are reducible, and so cannot appear in a minimal criminal.

Notice that this method generalizes Kempe's ideas. Instead of
removing just one country and colouring the rest of the map with
four colours, we now remove several countries at a time – in this
case, the countries inside or outside the ring. We then colour the
rest of the map with four colours, and match up the separate
colourings on the ring to give a colouring for the entire map.

The corresponding problem for rings of four countries is a
little more difficult because such a ring can be coloured with two,
three or four colours, and it may not be easy to match up the

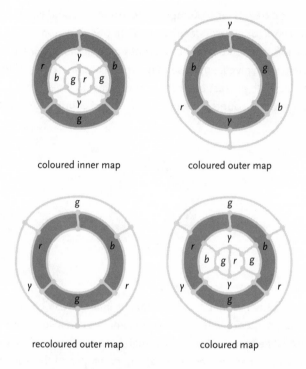

coloured inner map   coloured outer map

recoloured outer map   coloured map

colourings. The following diagrams show ring colourings with two
and three colours that cannot be matched up directly:

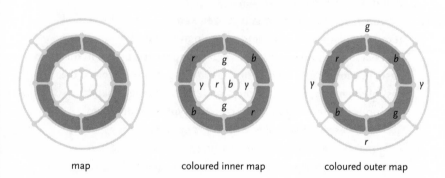

map   coloured inner map   coloured outer map

However, by switching the colours in appropriate two-coloured Kempe-chains, Birkhoff showed that these difficulties can always be overcome. He deduced that rings of four countries are reducible, and so cannot appear in a minimal criminal.

Birkhoff then extended his arguments to rings of five countries, and succeeded in every case except one: a single pentagon surrounded by a ring of five countries – the case that had defeated Kempe. Birkhoff's arguments also extended to certain rings of six countries, but these proved to be much more problematic. While carefully analysing them, Birkhoff became increasingly uncertain of the status of the four-colour problem, and in his paper he concluded that any of the following possibilities seemed plausible:

1. There exist maps which can not be colored in four colors, a leading feature of the simplest one of them perhaps being a ring of six regions with more than three regions on each side. By the method of reduction one will always be led either to a coloring of the given map, or to one or more maps that can not be colored.
2. All maps can be colored in four colors and a set of reducible rings can be found, one of which exists in every map.
3. All maps can be colored in four colors, but only by means of reductions of a more extensive character, applicable to sets of regions bounded by any number of rings.

Over thirty years later, in an important but highly technical paper, Arthur Bernhart of the University of Oklahoma managed to complete Birkhoff's work on rings of six countries. The story is told that, shortly after Bernhart married, his new wife encountered Mrs Birkhoff at a mathematics meeting. Mrs Birkhoff quizzed Mrs Bernhart:

Tell me, did *your* husband make you draw maps for him to color on your honeymoon, as mine did?

Whatever the answer, four colours clearly ran in the blood, as the Bernharts' son Frank also became a well-known writer on the four-colour problem.

In the next section we shall outline Birkhoff's argument for showing that a particular six-ring configuration of countries – the *Birkhoff diamond* – is reducible. This is an important configuration, once described as enjoying 'as much renown in graph theory as the Kohinoor diamond does in fictional criminal mysteries'. After this the floodgates were open, and mathematicians on both sides of the Atlantic began to develop Birkhoff's ideas, generating reducible configurations galore.

The four-colour problem also became a fashionable topic for research degrees, and a number of students obtained doctoral degrees for dissertations on map colouring. One of these students, whom we mentioned earlier, was Philip Franklin. In his Princeton thesis 'On the map color problem' he showed that each of the following configurations is reducible, and so cannot appear in a minimal criminal:

a pentagon in contact with three pentagons and one hexagon;
a pentagon surrounded by two pentagons and three hexagons;
a hexagon surrounded by four pentagons and two hexagons;
and any n-sided polygon in contact with $n - 1$ pentagons.

By applying the counting formula, he was then able to deduce that every map with up to 25 countries can be coloured with four colours, and therefore that a minimal criminal must have at least 26 countries.

Another young researcher was Alfred Errera, who wrote his dissertation 'Du coloriage des cartes' at the University of Brussels. Errera extended Franklin's results, proving in particular that a minimal criminal must contain at least thirteen pentagons and that it cannot contain only pentagons and hexagons.

Later mathematicians continued this work, obtaining further reducible configurations and using the counting formula to prove the four-colour theorem for maps with more and more countries. In particular, Clarence Reynolds of West Virginia showed in 1926 that four colours are sufficient for all maps with up to 27 countries. Franklin then increased this number to 31 in 1938, and two years later C. E. Winn, of the Egyptian University in Cairo, increased it to 35, where it stuck for a quarter of a century. Thus, a hundred years after De Morgan's original letter of 1852, it had been proved that all maps with up to 35 countries can be coloured with four colours – but there was still a long way to go.

We conclude this general discussion of reducible configurations by remarking that several of the above ideas had surfaced much earlier, in a most unlikely place. Like so many others, the French writer and poet Paul Valéry was fascinated by the four-colour problem, and his diaries for 1902 were later found to contain a dozen pages of substantial work on configurations of countries that were later to be explored by Birkhoff, Franklin and Winn.

COLOURING DIAMONDS

In this section we illustrate Birkhoff's methods by proving that the so-called Birkhoff diamond of four pentagons (shown overleaf) is reducible, and so cannot appear in a minimal criminal.

Suppose that we have a minimal criminal that contains the Birkhoff diamond. Removing the diamond yields a new map with fewer countries. We assume that this new map can be coloured with four colours, and try to extend the colouring to the pentagons in the diamond.

If the countries in the ring surrounding the diamond are numbered 1, 2, 3, 4, 5 and 6, as shown, it turns out that there are essentially 31 different ways in which they can be coloured with the

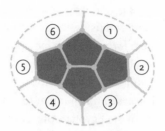

Birkhoff diamond

four colours *red*, *green*, *blue* and *yellow*. These colourings are as follows (the reason for the asterisks is given below):

rgrgrg   rgrbrg*   rgrbgy*   rgbrgy   rgbryb   rgbgbg*   rgbyrg   rgbygy*
rgrgrb*   rgrbrb   rgrbyg*   rgbrbg* rgbgrg* rgbgby   rgbyrb   rgbybg*
rgrgbg   rgrbry   rgrbyb*   rgbrby   rgbgrb* rgbgyg   rgbyry* rgbyby*
rgrgby*   rgrbgb*   rgrbgb   rgbryg   rgbgry* rgbgyb   rgbygb

Notice that we cannot include colourings such as *rgygbr* in which the two countries coloured *red* appear together, and we have also omitted colourings such as *rgrgry*, since this is essentially the same colouring as *rgrgrb* (on recolouring the final *yellow* as *blue*).

   Consider the colouring *rgrgrb*. This can be extended directly to the diamond, as shown below, and for this reason it is called a *good colouring*. In the same way, all the asterisked colourings above are good colourings: try some of them and see!

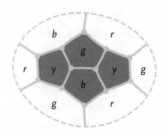

The colouring *rgrbrb* is not a good colouring, but by using Kempe-chain changes of colour that interchange the colours *red* and *yellow*, or *green* and *blue*, it can be converted into one of the good colourings *rgrgrb*, *rgrbrg* or *rgrbyb*. For example, if there is a *red–yellow* chain connecting countries 3 and 5, then we can interchange the colours in the *blue–green* chain containing country 4 so as to recolour country 4 *green*. Similarly, if there is a *red–yellow* chain connecting countries 1 and 5, then we can interchange the colours in the *blue–green* chain containing country 6 so as to recolour country 6 *green*. However, if there is no *red–yellow* chain connecting countries 3 and 5, or 1 and 5, then we can interchange the colours in the *red–yellow* chain containing country 5 so as to recolour country 5 *yellow*. (These three situations are illustrated below.) Thus, the colouring *rgrbrb* can be converted into a good colouring.

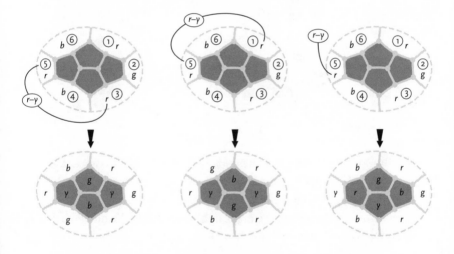

The colouring *rgrbry* is also not a good colouring, but by using Kempe-chain changes that interchange the colours *red* and *green*, or *blue* and *yellow*, it can be converted into either *rgrbgy* (which is good) or into *rgrbrb* (which can be made good, as we have just seen). This is because, if there is a *blue–yellow* chain connecting countries 4 and 6, then we can interchange the colours in the *red–green* chain containing country 5 so as to recolour country 5 *green*. However, if there is no *blue–yellow* chain connecting countries 4 and 6, we can interchange the colours in the *blue–yellow* chain containing country 6 so as to recolour country 6 *blue*. (These two situations are illustrated below.) Thus, the colouring *rgrbry* can also be converted into a good colouring.

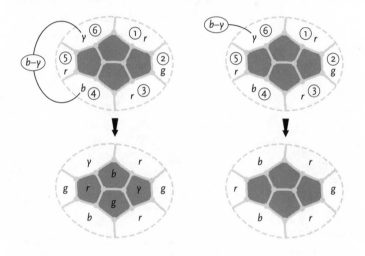

It turns out that each of the thirty-one possible colourings of the ring is either a good colouring, or can be converted into a good colouring by suitable Kempe-chain changes of colour. Thus, all thirty-one colourings of the ring can be extended to the diamond, so the Birkhoff diamond is reducible.

In fact, we do not have to consider all thirty-one colourings. If we modify the map by removing five of the boundary lines, as illustrated below, then we obtain a new map with fewer countries which can therefore be coloured with four colours:

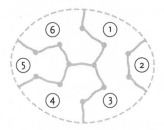

This modification corresponds to eliminating all those colourings in which countries 1 and 3 have different colours or countries 4 and 6 have the same colour. The effect of this is to eliminate all of the thirty-one colourings except *rgrgrb*, *rgrgby*, *rgrbrg*, *rgrbgy*, *rgrbyg* and *rgrbry*. The first five of these are good colourings, and the last can be converted into a good colouring, as we saw above. Thus, we can colour the new map with four colours, so the configuration is reducible.

Heinrich Heesch introduced the term *D-reducible* for configurations of countries for which every colouring of the surrounding ring of countries is a good colouring, or can be converted into a good colouring by a succession of Kempe-chain changes of colour. Thus, a digon, a triangle and a square are all *D*-reducible configurations, as shown by Kempe, and so is the Birkhoff diamond, as we have just seen. Heesch also used the term *C-reducible* for configurations that can be proved reducible after they have been modified in some way, as illustrated above. The concepts of *D*-reducible and *C*-reducible configurations will reappear in Chapters 9 and 10.

HOW MANY WAYS?

Birkhoff once remarked that almost every great mathematician
had worked on the four-colour problem at one time or another,
and in one of his last papers, a 97-page study of map colouring
written with D. C. Lewis and published posthumously, he classified all previous investigations into two types: *qualitative
approaches* and *quantitative approaches.*

  In the qualitative approach, the aim is to show that all maps of
a certain type can be coloured with four colours. Here, Kempe
chains play an important role, and the results for reducible configurations obtained by Birkhoff, Franklin, Errera, Reynolds and
Winn can be regarded among the successes of this approach. In
the quantitative approach we can use any number of colours, and
we want to find the number of ways in which a given map can be
coloured with these colours. Our object is to show that if four
colours are available, then the number of ways in which the map
can be coloured is a positive number. The quantitative approach
was introduced by Birkhoff while he was still at Princeton University, and his seminal paper was published in the *Annals of Mathematics* in 1912, the year before his paper on reducible
configurations and the Birkhoff diamond.

  To see what is involved, we start with this simple map:

  Birkhoff took the number of colours to be λ (lambda), where λ is
any number larger than 2. Then country *A* can be coloured with
any of these λ colours. Since country *B* is a neighbour of country *A*,

it can then be coloured with any of the remaining $\lambda - 1$ colours. Finally, since countries *C* and *D* are both neighbours of countries *A* and *B*, but not of each other, they can each be coloured with any of the remaining $\lambda - 2$ colours. Thus, the total number of ways of colouring all the countries of this map is

$$\lambda \times (\lambda - 1) \times (\lambda - 2)^2.$$

For example, if we have four colours ($\lambda = 4$), then the number of ways of colouring this map is $4 \times 3 \times 2^2 = 48$, and if we have ten colours ($\lambda = 10$), then the number of ways of colouring it is $10 \times 9 \times 8^2 = 5760$.

Birkhoff used the symbol $P(\lambda)$ to denote the number of ways of colouring the map with λ colours; thus, for the above map,

$$P(\lambda) = \lambda \times (\lambda - 1) \times (\lambda - 2)^2,$$

which we can multiply out to give

$$P(\lambda) = \lambda^4 - 5\lambda^3 + 8\lambda^2 - 4\lambda.$$

As a check we can substitute $\lambda = 4$ into this formula, giving

$$P(4) = 4^4 - (5 \times 4^3) + (8 \times 4^2) - (4 \times 4)$$
$$= 256 - (5 \times 64) + (8 \times 16) - 16 = 48,$$

as before.

The expression $P(\lambda)$, involving multiples of powers of λ, is called a *polynomial in* λ, and the numbers 1, -5, 8 and -4 preceding these powers are its *coefficients*. Birkhoff proved that the number of ways of colouring any map with λ colours is always a polynomial in λ, which he called the *chromatic polynomial* of the map, and he derived expressions for its coefficients in terms of properties of the map. Notice that if $P(4)$ is a positive number, then the map can be coloured with four colours.

Another result that Birkhoff proved was pointed out to him by his Harvard University doctoral student Hassler Whitney. If we look at the coefficients 1, −5, 8 and −4 of the chromatic polynomial above, we notice that these coefficients alternate in sign between positive and negative. It turns out that this is always the case:

> for any map, the coefficients of the chromatic polynomial alternate in sign.

In a paper presented to the American Mathematical Society on 25 October 1930, Whitney proved this result and then related his conclusions to properties of the map.

Birkhoff always hoped that he could solve the four-colour problem by investigating the properties of these chromatic polynomials, $P(\lambda)$. He wrote four papers about them, including the lengthy and very technical paper with Lewis mentioned earlier. Among the results that he obtained was the inequality

$$P(\lambda) \geq \lambda \times (\lambda - 1) \times (\lambda - 2) \times (\lambda - 3)^{n-3},$$

which is true for any map with *n* countries, and where λ is any positive integer *except* 4. If he could also have proved this result for $\lambda = 4$, then he would have shown that

$$P(4) \geq 4 \times 3 \times 2 \times 1^{n-3} = 24,$$

which is a positive number. He would therefore have proved that every map can be coloured with four colours (in at least 24 ways, in fact), thereby solving the four-colour problem.

Since Birkhoff's death there has been much further work on chromatic polynomials. They have been calculated for large numbers of maps, and their properties have been investigated in great detail. We conclude this chapter with a bizarre result due to Bill Tutte and his colleagues, derived in the late 1960s.

If a map is large, then we almost certainly need four colours to colour its countries. This means that it cannot be coloured with one, two or three colours, and so $P(1)$, $P(2)$ and $P(3)$ are all 0, whereas $P(4)$ is greater than 0. Now, are there any numbers *x*,

other than 1, 2 and 3, for which $P(x) = 0$? If there are, then
investigating them may give us an insight into the behaviour of
chromatic polynomials in general.

Tutte's result is related to the *golden ratio*, given by $\frac{1}{2}(1 + \sqrt{5})$
$= 1.618034\ldots$ This number, sometimes denoted by the Greek
letter τ (tau), appears throughout mathematics. It is the ratio of a
diagonal and a side of a regular pentagon, and a 'golden rectangle'
with sides in this ratio is sometimes considered to have the most
pleasing shape – neither too thin nor too fat:

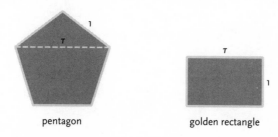

pentagon   golden rectangle

The golden ratio τ has interesting numerical properties. For
example, if we calculate its reciprocal and its square, we get exactly
the same pattern of decimals:

$\tau = 1.618034\ldots$, $1/\tau = 0.618034\ldots$ and $\tau^2 = 2.618034\ldots$

These results can easily be deduced from the quadratic equation
$x^2 = x + 1$, whose solutions are τ and $-1/\tau$.

In 1969, Gerald Berman and Bill Tutte noticed that, for large
cubic maps, the value of the chromatic polynomial $P(x)$ tends to
lie extremely close to 0 when $x = \tau^2$, usually with an agreement to
several decimal places. In the following year, Tutte supported
these observations theoretically by proving that, when $x = \tau^2$, the
value of $P(x)$ can never exceed τ^{5-n}, where n is the number of
countries in the map. For example, for cubic maps with twenty
countries, the value of $P(\tau^2)$ cannot exceed $\tau^{5-20} = \tau^{-15}$, which is

about 0.0007, while for cubic maps with thirty countries, the value of $P(\tau^2)$ cannot exceed $\tau^{5-30} = \tau^{-25}$, which is about 0.000006. Thirty years later, it is still not clear what implications these results have for the four-colour problem.

A new dawn breaks

By the middle of the twentieth century, much progress had been made on solving the four-colour problem. Only three papers had appeared on unavoidable sets of configurations, by Wernicke (1904), Franklin (1922) and Lebesgue (1940), but dozens of reducible configurations were being discovered, following on from the work of Birkhoff. Also, as we saw in Chapter 8, Franklin, Winn and others had demonstrated that if the four-colour theorem were false, then any minimal criminal would be complicated, containing more than thirty-five countries.

Meanwhile, large numbers of papers were appearing in the closely related area of graph theory – the study of 'linkages', mentioned by Kempe and Tait and later developed by Hassler Whitney and others. Graph theory was proving to be of great importance through its applications, particularly in a wide range of network problems on which great advances had been made during the 1950s. Many mathematicians were being attracted to the subject as an area of interest in its own right, largely through the appearance of important textbooks. These included the first major book on the subject, Dénes König's *Theorie der endlichen und unendlichen Graphen* ('Theory of finite and infinite graphs'), which appeared in 1936, and later textbooks on graph theory by Claude

Berge in France and by Oystein Ore, Robert Busacker and Thomas Saaty, and Frank Harary in the United States.

The 1960s also proved to be an exciting time for map colouring. In 1967 the first major book devoted exclusively to map colouring, Oystein Ore's authoritative and influential *The Four-Color Problem*, was published. The next year, Ore and his research student, Joel Stemple, extended the earlier methods of Franklin and Winn to prove that all maps with up to forty countries can be coloured with four colours. They needed to consider so many special cases in their proof that the full details could not be published but had to be deposited with the Mathematics Department Library at Yale University. (When the general proof finally appeared, it was also criticized for not being publishable in full.)

DOUGHNUTS AND TRAFFIC COPS

One undoubted success story around this time was the proof in 1968 of the Heawood conjecture by Gerhard Ringel from Germany and Ted Youngs of California. As we saw in Chapter 7, Percy Heawood had proved that every map drawn on the surface of an h-holed torus can be coloured with $H(h) = [\frac{1}{2}(7 + \sqrt{1 + 48h})]$ colours, but he failed to prove that there are maps on an h-holed torus that actually require this number of colours, when h is larger than 1. As Gerhard Ringel later wryly commented:

In 1890 P. J. Heawood published a formula which he called the Map Colour Theorem. But he forgot to prove it. Therefore the world of mathematicians called it the Heawood Conjecture. In 1968 the formula was proven and therefore again called the Map Color Theorem.

The proof was a veritable tour de force. Just as Möbius's problem of the five princes is equivalent to the problem of connecting five palaces with non-crossing roads (see Chapter 2), so Heffter in 1891 showed that the Heawood conjecture is equivalent to a problem about the connecting of n points with non-crossing lines on a torus with a certain number of holes. This certain number involved the fraction $\frac{1}{12}$, and the denominator 12 turned out to be highly significant: in fact, the eventual proof of the Heawood conjecture split into twelve completely separate cases, depending on the remainder when n is divided by 12.

By the summer of 1967, all but three of these cases had been settled, and Ted Youngs invited Ringel to visit California for the academic year 1967/68 to work on them. They struggled for several months, and eventually the proof was completed. Great rejoicing!

Ted Youngs (left) and Gerhard Ringel in 1968.

Solving map-colouring problems can sometimes yield unexpected benefits. Shortly after the proof of the Heawood conjecture had been announced, Ringel was driving along the California expressway and was stopped by a traffic cop for a minor traffic violation. As soon as he found that the culprit's name was 'Ringel', the cop asked, 'Are you the one that solved the Heawood conjecture?' Ringel, surprised, admitted that he was, and was duly let off with only a warning. As it happened, the cop's son had been in Ted Youngs' calculus class when the proof of the Heawood conjecture had been announced.

Before leaving the Heawood conjecture, notice that if we substitute $h = 0$ into Heawood's formula, then we obtain the correct answer for the number of colours required for maps on 'the torus with no holes' – in other words, the sphere:

$$H(0) = [\tfrac{1}{2}(7 + \sqrt{1})] = [4] = 4.$$

Unfortunately, we cannot deduce the four-colour theorem from the proof of the Heawood conjecture, which applies only when h is a positive number.

HEINRICH HEESCH

In the 1960s, much of the work on the four-colour problem was still piecemeal. Attempts to find unavoidable sets and reducible configurations were largely independent of one another, and a co-ordinated search for the holy grail of an unavoidable set of reducible configurations had yet to meet with any success. The credit for advocating such a search goes to Heinrich Heesch, whose contributions would pave the way for the eventual solution of the problem by Kenneth Appel and Wolfgang Haken in the 1970s.

Heesch's early career in mathematics was spectacular. After graduating in both mathematics and music from Munich, he

Heinrich Heesch (1906–95)

obtained a doctorate in mathematics from the University of Zurich for a dissertation on problems related to symmetry in three-dimensional geometry. He then moved to Göttingen, the home of such distinguished mathematicians as David Hilbert, Richard Courant and Hermann Weyl, where he became Weyl's assistant, studying the geometry of crystals. While in Göttingen, he gained some notoriety for contributing to the solution of one of the celebrated *Hilbert problems*.

David Hilbert, arguably the greatest mathematician of his time, gave a lecture at the Second International Congress of Mathematics, held in Paris in 1900, in which he set out the twenty-three mathematical problems that he hoped to see solved during the twentieth century. Several of his problems have indeed been settled, while others remain unanswered to this day. The four-colour problem, lying outside the mainstream of mathematics, was not one of these twenty-three problems, but the *regular parquet problem*, on the construction of a particular type of tiling of the plane, was part of Hilbert's Problem 18. Heesch solved this problem in 1932 by constructing a number of tilings that can be used to cover the plane according to the rules laid down in the problem. One of his tiling patterns was later incorporated into the ceiling of Göttingen's library.

The mid-1930s were difficult years for Heesch, as they were for many other German academics and intellectuals. He was unsympathetic towards the National Socialists, objecting in particular to the Nazi work camps that were set up for aspiring professors, and soon found himself without university employment. For twenty years Heesch taught mathematics and music in various schools and worked on industrial problems that involved tiling patterns, while continuing with his researches into mathematics. He eventually secured a teaching position at the Technical University of Hanover.

Heesch's interest in the four-colour problem dates from around

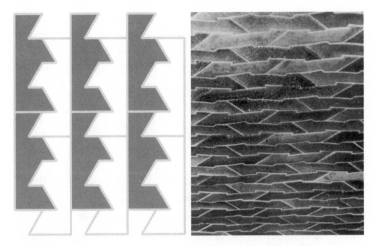

Heesch tiling   The ceiling of Göttingen's library.

1935. One of his friends, Ernst Witt, believed that he had found a
solution to the four-colour problem, and the two men visited
Richard Courant to show him the solution. Courant was about to
leave for a train journey to Berlin, so Heesch and Witt bought train
tickets and accompanied him on his journey. Failing to convince
Courant that the proof was correct, and feeling rather deflated,
they caught the train back to Göttingen; on the return journey,
Heesch found an error in Witt's solution.

While he was working on the four-colour problem, Heesch
gradually came to believe that the right approach was to seek an
unavoidable set of reducible configurations. But he feared that
such a set would be very large, possibly containing up to ten thousand configurations.

In order to produce unavoidable sets, he invented the method
of discharging, which we outlined in Chapter 8. He also developed
the uncanny knack of being able to look at a configuration and say
whether it was reducible – with at least 80 per cent accuracy. As
Wolfgang Haken later remarked:

What fascinated me most was that Heesch looked at the configuration, and he either said, 'No, there is no chance. That cannot be reducible', or he said, 'But this one: that is certainly reducible'. And I asked him, 'How do you know? How can you tell?' 'Well, I need two hours' computer time . . .'

WOLFGANG HAKEN

In the late 1940s, Heesch presented his findings publicly for the first time, in lectures at the University of Hamburg and in his home town of Kiel, in Schleswig-Holstein. In these lectures Heesch expounded on his belief that there exists an unavoidable set of reducible configurations, that these configurations should not be particularly large, but that there is likely to be a very large number of them.

One student who attended Heesch's 1948 lecture at the University of Kiel was the young Wolfgang Haken. Haken had entered the university as its youngest student and was studying mathematics, philosophy and physics. Haken recalls hearing Heesch's talk, much of which he did not understand at the time, but remembers Heesch mentioning that there might be some ten thousand cases to be investigated, and that five hundred configurations had already been checked – at an average rate of about one per day. Heesch seemed optimistic about dealing with the remaining nine and a half thousand.

Some of the most stimulating lectures that Haken attended at Kiel were those on topology given by the only mathematics professor, Karl Heinrich Weise, who had worked on the four-colour problem. In these lectures Weise described three long-standing unsolved problems. First was the *knot problem*, of determining whether a given tangle of string in three dimensions contains a

knot – for example, the first tangle below can be unknotted, while
the second one must always remain knotted.

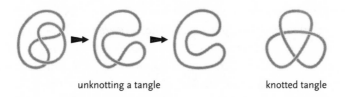

unknotting a tangle   knotted tangle

The second problem was the *Poincaré conjecture*, which remains
unsolved to this day, and concerns the classification of spheres in
four-dimensional space. The third was the four-colour problem.

Like David facing Goliath, Haken took on all three problems, in
each case using 'very elementary means on which everyone else
had given up'. The first Goliath he slew. The second he attacked
vigorously, rendering the giant 99 per cent vanquished, but his vic-
tim recovered. The third Goliath he battled with for several years,
eventually replacing the stone in his sling by an appel, as we shall
see in Chapter 10.

In his doctoral dissertation, written under the supervision of Professor Weise, Haken worked on three-dimensional topology and
obtained some partial results on the knot problem. Subsequently
he succeeded in solving the problem in full – a great achievement
– and announced his results at the 1954 International Congress of
Mathematicians in Amsterdam, where he was urged to write up
the details of the proof. But by this time, unable to secure a university position in Kiel or elsewhere, he was working in Munich as a
physicist for Siemens, the electrical engineering company. It took
Haken four years to find the time to write out the two hundred
pages of his proof in intricate detail, and it was to be a further
three years before his solution was published. It appeared in the
journal *Acta Mathematica* in 1961.

Haken's work on the knot problem so impressed Bill Boone, a

logician at the University of Illinois at Urbana-Champaign, that Haken was invited to Illinois as a visiting professor. Boone's interest in the problem had been sparked by Kurt Gödel's incompleteness theorem, published back in 1931, which asserted that in any mathematical system, however complicated, there are always some problems that cannot be solved within the system. Because the knot problem and the four-colour problem stubbornly resisted all attempts at solution, the feeling grew in the 1950s that both problems were 'unsolvable', and that one could never know whether their statements were true or false. In the event, Haken settled the first of these problems and, with Appel and others, solved the second.

Shortly after arriving in Illinois, Haken gave several lectures on his researches into the knot problem. His painstaking approach to solving mathematical problems prompted one of his colleagues to observe:

Mathematicians usually know when they have gotten too deep into the forest to proceed any further. That is the time Haken takes out his penknife and cuts down the trees one at a time.

After spending a couple of years at the Institute for Advanced Study in Princeton, Haken returned to take up a permanent position at the University of Illinois. There he was able to continue his work on the Poincaré conjecture, which he reduced to the consideration of 200 particular cases. As he recalls, he then worked systematically and painstakingly through these cases until no fewer than 198 had been successfully completed. For thirteen years he struggled with the final two cases of this 'man-eating' problem, before abandoning them to work on the four-colour problem.

ENTER THE COMPUTER

In 1967 Haken contacted Heesch. With his experience of the Poincaré problem, in which he had failed to crack just two out of two hundred cases, Haken feared that the same thing might have happened to Heesch with his ten thousand configurations, and that Heesch might have given up. But no: Heesch was still working on the problem, and by this time he had invented the method of discharging and had discovered thousands of reducible configurations.

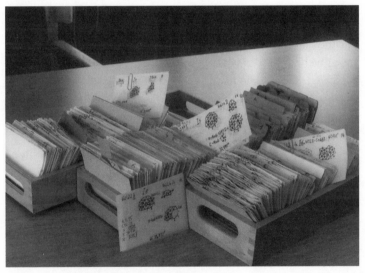

Heesch's box files of reducible configurations.

Heesch's aim was to 'systematize' the ideas in Birkhoff's 1913 paper by developing an approach for generating reducible configurations. He first looked at the simplest type of reducible configuration, which he called *D-reducible* – remember from Chapter 8 that these are configurations for which every colouring of the surrounding ring of countries can be extended to the countries in the centre, either directly or by using Kempe-type interchanges of colour. He also introduced the term *C-reducible* for a configuration that can be proved to be reducible after it has been modified in such a way as to restrict the number of ring colourings that have to be considered. If a configuration was not *D*-reducible, Heesch could frequently see how to modify it so as to determine whether it was *C*-reducible. By this means he was able to whittle the problem down to a smaller number of possible colourings, as we did in our discussion of the Birkhoff diamond.

Haken invited Heesch to the University of Illinois to give a lecture, and raised the question of whether computers could be helpful in the examination of large numbers of configurations. In fact, this thought had already occurred to Heesch, and in the mid-1960s he had enlisted the help of Karl Dürre, a mathematics graduate from Hanover who had become a secondary school teacher. Dürre was able to develop a method for testing *D*-reducibility which was sufficiently routine and algorithmic to be implemented on a computer, even though it might take a long time. By November 1965, using the programming language Algol 60 on the University of Hanover's CDC 1604A computer, Dürre was able to confirm that the Birkhoff diamond is *D*-reducible, and soon established the *D*-reducibility of many more configurations of increasing complexity.

The complexity of a configuration is measured by its *ring-size*, which is the number of countries surrounding the configuration. For the Birkhoff diamond (opposite, top), which has ring-size 6, there are 31 essentially different colourings of the countries in the surrounding ring of six countries, as we saw in Chapter 8.

Unfortunately, the number of different colourings increases rapidly as the ring-size increases:

ring-size	6	7	8	9	10	11	12	13	14
colourings	31	91	274	820	2461	7381	22144	64430	199291

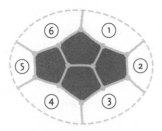

Birkhoff diamond

For example, the configuration of eight countries shown below has ring-size 14, and we have to consider no fewer than 199291 essentially different colourings of the countries in the surrounding ring. Informal calculations indicated that to solve the four-colour problem might require the examination of configurations with ring-size 18, for which the number of colourings of the surrounding ring exceeds 16 million.

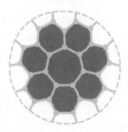

Heesch and Dürre found that the time taken to analyse a configuration increased rapidly as the ring-size increased. On their computer a typical configuration with ring-size 12 might take six hours to analyse, while some configurations with ring-size 13 took anything between sixteen and sixty-one hours, and those of ring-size 14 were way over the horizon. Indeed, they estimated that to verify all ten thousand cases they might need anything from three thousand to fifty thousand hours of computer time, which was not a realistic proposition on the Hanover computer – or indeed on any computer of the time.

One useful simplification that Dürre discovered was a more efficient way of storing the colourings of the surrounding ring. For a configuration of ring-size 14, storing the colours of the fourteen countries in the surrounding ring would require 28 bits, using two bits to specify the colour (00, 01, 10 or 11) of each country. Since there are almost two hundred thousand possible colourings for a configuration of ring-size 14, over 5 million bits would be needed for each such configuration. Dürre developed a brilliant technique that enabled him to assign a single bit to each colouring, thereby reducing the total amount of computer time for each configuration by a factor of 28 – a considerable help. Although most configurations of ring-size 14 were still too large to be handled, many smaller configurations could now be checked for the first time.

By this time, mathematicians could see that any solution of the four-colour problem along these lines would be complicated. In the 1960s, Edward F. Moore of the University of Wisconsin developed the remarkable knack of constructing large and intricate maps that contained no known reducible configurations, thereby hoping to find a map that required five colours. The following diagram shows part of a map he devised that contains no reducible configurations with ring-size 11 or less: here the left-hand and right-hand sides of the map must be joined up, so that the top and bottom countries are both 9-sided. The existence of this map shows that any unavoidable set of reducible configurations must necessarily contain at least one reducible configuration of ring-size 12 or greater.

Although it looked as if configurations of ring-size up to 18 might be necessary, Appel and Haken's solution would be based on configurations with ring-size 14 and less. Had it required configurations of ring-size 15, Appel and Haken would almost certainly not have been able to complete their solution when they did.

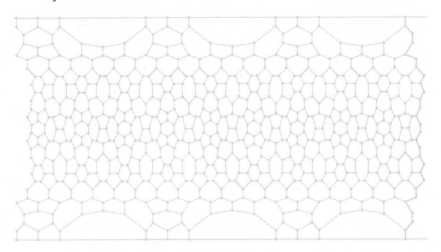

By this time, almost everyone working on the four-colour problem was using the 'dual formulation', introduced by Kempe, of colouring the points of the corresponding graph or linkage. (Unlike Heesch and his successors, we avoid a sudden change of gear at this point by continuing to present our story in terms of maps.) In particular, Heesch devised a useful notation, which soon became widely used, for representing each point by an appropriate 'blob' to make it easily distinguishable. Here are four of Heesch's symbols followed by an example of a Heesch drawing and a corresponding configuration of countries.

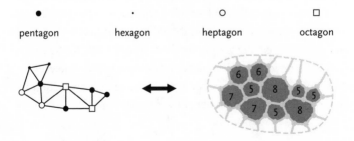

Although Heesch lectured widely on his work, it remained unpublished until 1969, when he produced a paperback in German setting out the method of discharging and his many other contributions. Oystein Ore, in his 1967 book on the four-colour problem, had made no mention of Heesch's work, of which he was presumably unaware. However, Heesch was pleased to learn about several new reducible configurations from Ore's book.

COLOURING HORSESHOES

It soon became clear that the Hanover computer was not powerful enough to carry out the work required of it. Haken tried to help Heesch and Dürre by bidding for time on the large new parallel supercomputer at the University of Illinois, the ILLIAC-IV, whose construction was nearing completion, but it was not yet ready for use. Eventually, John Pasta, head of the university's computer science department, referred Heesch and Dürre to his friend Yoshio Shimamoto, Director of the Computer Center at the Atomic Energy Commission's Brookhaven Laboratory in Upton, Long Island. At this laboratory there was a Cray Control Data 6600, the most powerful computer of its day.

Fortunately, Shimamoto had always been a devotee of the four-colour problem, and as director he was allowed to use up to 10 per cent of his time on the computer for his own purposes. He was strongly supportive of Heesch's approach to the problem, and generously invited Heesch and Dürre to visit Brookhaven and continue their reducibility testing on the Cray machine. In the event, Heesch paid two extended visits, including a stay of one year, while Dürre remained there for almost two years, completing a doctoral thesis on reducibility at the same time.

A page of Heesch's diagrams.

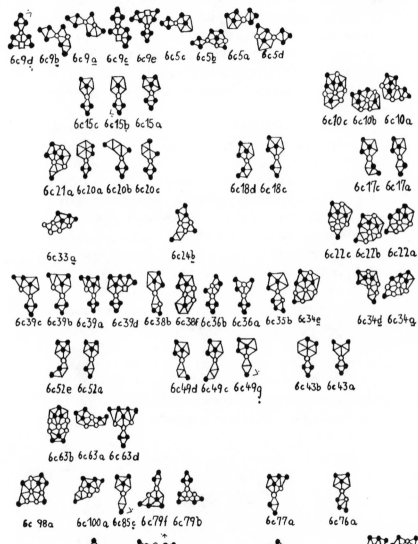

△ **6c**

6c9d 6c9b 6c9a 6c9c 6c9e 6c5c 6c5b 6c5a 6c5d

6c15c 6c15b 6c15a

6c10c 6c10b 6c10a

6c21a 6c20a 6c20b 6c20c

6c18d 6c18c

6c17c 6c17a

6c33a

6c24b

6c22c 6c22b 6c22a

6c39c 6c39b 6c39a 6c39d 6c38b 6c38f 6c36b 6c36a 6c35b 6c34e

6c34d 6c34g

6c52e 6c52a

6c49d 6c49c 6c49g

6c43b 6c43a

6c63b 6c63a 6c63d

6c98a

6c100a 6c85c 6c79f 6c79b

6c77a

6c76a

6c125a

6c118a

6c110a 6c109a

6c101b 6c101a

At Brookhaven, Dürre first had to convert his programs from Algol to Fortran, but once this was done progress was quickly made. A configuration with ring-size 13, with 66430 colourings to check, was not excessively large for the Cray computer, and configurations with ring-size 14 could be tested for the first time. Eventually, Heesch and Dürre were able to confirm the D-reducibility of more than a thousand configurations of ring-size 14 or less.

Early in the process, Dürre discovered a small bug in the program, which meant that the first few configurations on the list were not absolutely reliable. Since each of these configurations would have taken several hours to rework, he postponed the task. In the meantime, he repaired the bug, leaving strong warnings about these early configurations – but these warnings were lost.

Meanwhile, Shimamoto was pursuing his own researches on the four-colour problem, using a different but related approach. Unlikely as it may seem, he was able to show that if he could find a single configuration with certain properties, and if this configuration were D-reducible, then the four-colour theorem would follow: the whole of the proof would depend on a single configuration!

Suddenly, on 30 September 1971, he found what he was seeking. Like the Bishop of London in the 1880s (see Chapter 6), Shimamoto was attending a rather boring meeting and he started to colour maps – eventually producing the following configuration

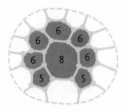

Shimamoto horseshoe

with ring-size 14 that became known as the *Shimamoto horseshoe* (even though it looks nothing like a horseshoe). Shimamoto had proved that if this horseshoe were *D*-reducible, then the four-colour problem was solved.

After the meeting, Shimamoto bumped into Heesch and Haken, who were visiting Brookhaven and were on their way to the cafeteria for lunch. Heesch recognized the horseshoe as one of his list of *D*-reducible configurations. Shimamoto, although obviously excited, was cautious and asked for the configuration to be rechecked for *D*-reducibility. Dürre came over from Germany to do the rechecking, but the original printouts were no longer available and so the whole task had to be run from scratch.

Excitement ran riot! Rumours about Shimamoto's solution were already spreading around the world. The day after the horseshoe's discovery, Haken was due to visit the Princeton Institute for

Heinrich Heesch, Yoshio Shimamoto and Karl Dürre at Brookhaven National Laboratory.

Advanced Study before returning to Illinois, and he asked Shimamoto if he could to tell people about the result. According to
Haken, Shimamoto said 'Yes', while Shimamoto later denied
doing so and became rather annoyed about Haken's pronounce-
ment, complaining that

I am getting a deluge of questions from many people and I am now
wondering about the wisdom of having you talk at Princeton and
Urbana before a clean manuscript had been prepared.

In the event, things did not work out as hoped. In the first computer run, the time taken exceeded the original run time by more
than an hour, and it was duly terminated. The second run took
even longer. Finally, in a third attempt, the super-fast Cray computer was allowed to run for a whole weekend, and after grinding
on for twenty-six excruciating hours, it confirmed that the horseshoe was, after all, *not* D-reducible. Whether the horseshoe was
one of Dürre's bugged configurations was unclear, but there was
certainly a sense of great disappointment all round.

Haken reworked Shimamoto's theoretical arguments and found
them completely correct. Hassler Whitney and Bill Tutte, the most
distinguished graph-theorists of their day, also analysed the Shimamoto approach in detail, and wrote a lengthy and influential paper
that began:

In October 1971 the combinatorial world was swept by the rumour
that the notorious Four Colour Problem had at last been solved . . .
with the help of a computer.

They then summarized the method:

Shimamoto, on the assumption that the Four Colour Conjecture
was false, showed that there must be a non-colourable map M
containing a configuration H [the horseshoe] that had already
passed the computer test for D-reducibility. He then arrived at a

contradiction by showing that the D-reducibility of H implied the 4-colourability of M . . . The burden of proof was not now on a few pages of close reasoning, but on a computer!

All along, Whitney and Tutte had felt that, if Shimamoto's proof were valid, a much simpler proof should be achievable by confining one's attention to a smaller part of M, and that 'this simpler proof would be so simple that its existence was incredible'. Since they found no essential flaw in Shimamoto's reasoning, they deduced that the computer result must be wrong. Moreover, they proved that not only was the horseshoe not D-reducible, but so also was *any* configuration produced by Shimamoto's method.

It seemed as though the four-colour problem had once again come to a dead end.

Success! . . .

In spite of the horseshoe episode, there were reasons for optimism. Heesch's approach to the four-colour problem was beginning to pay dividends, and in the hands of Appel and Haken would yield the eagerly desired goal of an unavoidable set of reducible configurations within the next five years.

In Chapter 8 we saw that Heinrich Heesch's method of discharging involved assigning an electrical charge of $6 - k$ to each country bounded by k edges, so that each pentagon receives a charge of 1, hexagons receive zero charge, and heptagons, octagons, etc., receive a negative charge. In 1970, Heesch sent Haken the results of a new discharging experiment in which the positive charge on each pentagon was now to be distributed *equally* among all neighbouring countries with negative charge. The result of this experiment, if applied to a general map, would yield about 8900 'bad' configurations extending up to ring-size 18, in which some countries would still have positive charge. This approach, called the *finitization of the four-colour problem*, reduced the problem to the consideration of these 8900 configurations, which Heesch proposed to work through one at a time.

Wolfgang Haken, however, was deeply pessimistic about having to deal with such a large number of configurations, especially since some of them were fairly big. By this time it had become a

Kenneth Appel and Wolfgang Haken contemplate their solution.

simple matter to test configurations of ring-size 11, since there are only 7381 different colourings of the surrounding ring of eleven countries, but each time the ring-size increased by one, the amount of computer time required seemed to expand by a factor of four, with a corresponding increase in storage requirement. Recalling that the difficult horseshoe configuration with ring-size 14 had taken many hours to run, Appel and Haken later observed:

Even if the average time required for examining 14-ring configurations was only twenty-five minutes, the factor of 4 to the fourth power in passing from 14- to 18-rings would imply that the average eighteen-ring configuration would require over a hundred hours of time and much more storage than was available on any existing computer. If there were a thousand configurations of ring-size 18, then the whole process would take over a hundred thousand hours, or about eleven years, on a fast computer.

A HEESCH–HAKEN PARTNERSHIP?

For some time, Haken had felt that the complexity of the problem would be substantially reduced if a better discharging method could be found. By focusing on maps without hexagons or heptagons, he succeeded in finding a much simpler procedure. Encouraged by this, he moved on to maps in general, and sent some of his results to Heesch. Heesch was impressed, and invited Haken to collaborate with him.

In 1971 Heesch sent Haken several of his unpublished results on reducible configurations. These included three 'obstacles to reducibility' which seemed to prevent a configuration from being reducible. Although it has never been proved that a configuration containing one of these obstacles cannot be reducible, no such reducible configuration has ever been found, so it seemed sen-

sible to exclude them from consideration. The general study of
such obstacles was soon developed by Whitney and Tutte in their
paper on Shimamoto's work, and later by Walter Stromquist, a
graduate student at Harvard University, who showed that
Heesch's trio are the three most important obstacles. Stromquist
would soon make a name for himself in the map-colouring world
by proving that the four-colour theorem is true for all maps with
up to 51 countries.

Heesch's three obstacles are these:

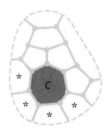 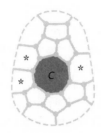 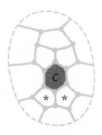

4–legger country   3–legger articulation country   hanging 5–5 pair

The first is a *4-legger country* – a country C that adjoins four consecutive countries (marked with stars) of the surrounding ring.
The second is a *3-legger articulation country*, a country C that
adjoins three countries of the surrounding ring that are not
mutual neighbours. The third is a *hanging 5–5 pair*, a pair of adjacent pentagons that adjoin a single country C inside the surrounding ring.

But by this time, Haken was beginning to change his approach
to the problem. Unlike everyone else, whose objective seemed to
be to collect reducible configurations by the hundreds and then
package them up into an unavoidable set, Haken's main line of
attack (which he later developed with Appel) was to aim directly
for an unavoidable set. This set was to contain only configurations
that were *likely* to be reducible – in particular, they should contain

none of the reduction obstacles – in order to avoid wasting
time checking configurations that would eventually be of no interest. Any configurations that subsequently proved not to be reducible could then be dealt with individually. As Haken later
commented:

If you want to improve something, you should not improve that
part which is already in good shape. The weakest point is always
the one you should improve. This is a very simple answer to why
we got it and not the others.

He also felt that it was inappropriate to spend a lot of expensive
computer time in checking the reducibility of configurations that
would be unlikely to appear in the eventual unavoidable set.

Thus, from this point on, Haken headed off in a different direction from everyone else, concentrating on the unavoidable set and
leaving the details of reducibility testing until much later. Heesch,
while initially sympathetic to these ideas, soon came to reject the
idea of a 'likely-to-be-reducible' configuration. Their working
relationship must have been further damaged by the horseshoe
episode, and Shimamoto had understandably decided that he no
longer wished to co-operate.

Back in Germany, Heesch was having immense problems
obtaining the necessary funding for the use of a powerful computer. As a relatively minor figure in the academic hierarchy, he
had little clout, and his proposal for funding did not receive the
consideration it deserved. The unsympathetic grant referee apparently showed little understanding of the problem or of Heesch's
approach to it, causing Shimamoto to protest:

The remarks made by this referee are so inane that they hardly
deserve any attention. It is evident that he knows nothing about the
problem, and by the bias shown in his comments, he should disqual-
ify himself from reviewing proposals involving the four-color

problem. It comes to me as something of a surprise that the Foundation itself did not see it appropriate to ignore his comments.

KENNETH APPEL

Meanwhile, Haken, with little knowledge of computing and with the Brookhaven machine no longer available, also considered setting the problem aside for a few years until more powerful computers became available to deal with the massive calculations that would clearly be necessary. He had talked to computer 'experts' who told him that his ideas could not be programmed, and during a lecture that he gave in Illinois on the horseshoe episode he exclaimed:

The computer experts have told me that it is not possible to go on like that. But right now I'm quitting. I consider this to be the point beyond which one cannot go without a computer.

Present at this lecture was Kenneth Appel. Appel had graduated from Queens' College, New York, before receiving his doctorate from the University of Michigan for a dissertation on the application of mathematical logic to some problems in algebra. He was an experienced computer programmer: he had learned programming at the University of Michigan, and gained more computing experience during a summer placement with Douglas Aircraft. After working at the Institute for Defense Analysis at Princeton for two years, he settled at the University of Illinois at Champaign-Urbana. Appel's computer skills were to prove invaluable in the solution of the four-colour problem.

After the lecture, Appel told Haken that he thought the 'experts' were talking nonsense – their opinion probably reflected their unwillingness to invest a great deal of time in something whose outcome was uncertain. Appel offered to work on the problem of implementing the discharging procedures saying, 'I don't know of anything involving computers that can't be done; some things just take longer than others. Why don't we take a shot at it?' Coincidentally, Thomas Osgood, a research student of Haken's, had just submitted his doctoral thesis on solving the four-colour problem for maps containing only pentagons, hexagons and octagons. Appel was one of the members of Osgood's thesis examination panel, so the collaboration might well prove beneficial to all concerned.

Haken was delighted to accept Appel's offer to take care of the computing side of things. They decided to concentrate their search on unavoidable sets, without taking time to check the configurations for reducibility. Instead, they focused on *geographically good* configurations – configurations that contain neither of Heesch's first two obstacles to reducibility, the 4-legger country and the 3-legger articulation country: such configurations can easily be identified by a computer – or, indeed, by hand. They would then check their configurations for reducibility once the entire set had been constructed.

Another useful rule-of-thumb they used to identify suitable configurations for inclusion was the *m-and-n rule*: if n is the ring-size of a configuration without obstacles, and m is the number of countries inside the surrounding ring, then the likelihood of reducibility depends on the relative sizes of n and m; in particular, if m is larger than $\frac{3}{2}n - 6$, then the configuration is almost certainly reducible. For example, the reducible Birkhoff diamond has ring-size 6 and 4 interior countries, and 4 is larger than $(\frac{3}{2} \times 6) - 6 = 3$; on the other hand, the Shimamoto horseshoe, which is not reducible, has ring-size 14 and 10 interior countries, and 10 is smaller than $(\frac{3}{2} \times 14) - 6 = 15$.

GETTING DOWN TO BUSINESS

When they started work in late 1972, Appel and Haken had no
clear idea of where they were heading. As Appel recalls, 'We
started with certain ideas and kept discovering that we had to
become more sophisticated to avoid being swamped by useless or
repetitive data.' However, their first exploratory computer runs
were already providing much useful information – in particular,
that geographically good configurations of reasonable size (with
ring-size 16 or less) tended to appear near most countries that
ended up (after discharging) with positive charge. However, the
computer output was enormous, with many configurations appearing over and over again. It would be necessary to keep these dupli-
cations under control if the eventual list was to be manageable.
Fortunately, since the computer runs had taken just a few hours,
Appel and Haken realized that they could experiment as frequently
as they needed to.

The necessary changes to the program were straightforward to
implement, and the second set of runs, a month later, saw a defi-
nite improvement. The printout was also much reduced in thick-
ness, and would eventually be down to a fraction of an inch. And
as the larger problems were overcome, so lesser ones began to
emerge.

From then on, every two weeks or so they modified the discharging algorithm or the computer program so that the program grew
while the output shrank. Their two-way dialogue with the computer continued as each successive problem was sorted out and
new ones arose. Six months of experimenting and improving their
procedures showed them that their method for producing a finite
unavoidable set of geographically good configurations in a reasonable amount of time was indeed feasible.

At this stage, they changed tack. They decided to prove

theoretically that their method would provide such an unavoidable set. To do this they would have to include every possible case, even if it was unlikely to occur in practice. This proved to be much more complicated than they had expected, taking them over a year to complete. But the eventual outcome, in the autumn of 1974, was a lengthy proof that an unavoidable set of geographically good configurations exists, together with an achievable method for constructing such a set. Shortly afterwards, Walter Stromquist, as part of his Harvard doctoral thesis, produced a shorter and more elegant proof of the existence of such a set.

Appel and Haken's next task was to decide how complicated the entire process would be. They decided to experiment with a particular case – maps that do not contain two adjacent pentagons. This was much simpler than the general case, and produced a set of just 47 geographically good reducible configurations of ring-size 16 or less, as shown opposite. This experiment enabled them to estimate that the general problem might be only about fifty times as bad as this, and on this basis they decided to proceed. In the event, their estimate turned out to be over-optimistic.

In early 1975 they introduced the third of Heesch's obstacles, the hanging 5–5 pair. Inevitably, this necessitated further changes in procedure, but was carried out successfully with only a doubling in the size of the unavoidable set. They also programmed the computer to search for sets of configurations with relatively small ring-size.

At this stage the computer started to think for itself, as Appel and Haken later recalled:

It would work out complex strategies based on all the tricks it had been 'taught' and often these approaches were far more clever than those we would have tried. Thus it began to teach us things about how to proceed that we never expected. In a sense it had surpassed its creators in some aspects of the 'intellectual' as well as the mechanical parts of the task.

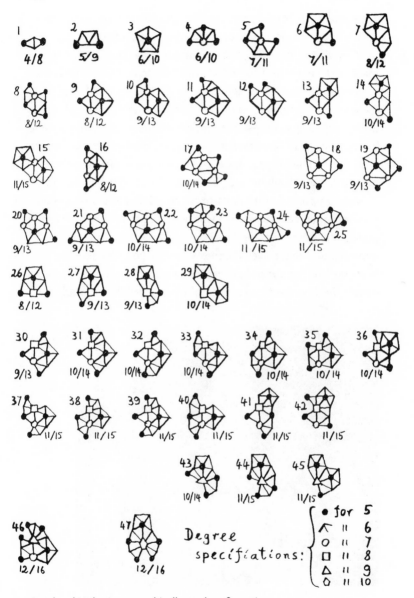

Appel and Haken's geographically good configurations.

As soon as it seemed probable that Appel and Haken would be able to find an obstacle-free unavoidable set of configurations, which were therefore likely to be reducible, it was time to start the massive detailed check for reducibility. Inevitably, there would be some rogue reducible configurations in the list, but they hoped that these would be relatively few in number. Also, with configurations that might go up to ring-size 16, and others that might cause trouble in other ways, they were expecting to have to devise some short cuts.

By the middle of 1974, Appel and Haken knew they would need help with the reducibility programs. Appel visited the University of Illinois computer science department to ask whether any graduate student would be interested in writing a thesis in the area of programming. Coincidentally, John Koch had just been scooped by a journal article that solved the thesis problem on which he had been working for a year, and was looking around for another topic. Koch's thesis project was set up so that its completion would not

John Koch and his wife with his doctoral certificate.

depend on the outcome of Appel and Haken's attack on the four-colour problem.

Koch was put to work on the C-reducibility of configurations of ring-size 11. As we saw with the Birkhoff diamond in Chapter 8, a C-reducible configuration is one that can be adapted so that the reducibility arguments go through more easily, but it is not always clear how this adaptation can be carried out. Appel and Haken were particularly interested in two types of modification that were relatively easy to implement, and Koch discovered that 90 per cent of the configurations of ring-size 11 were of these types. Agreeing that little would be gained by including the other 10 per cent, which would have required some difficult programming, they decided to concentrate exclusively on the simple modifications. Koch devised an elegant and highly efficient method for testing the C-reducibility of these configurations of ring-size 11, and Appel subsequently extended it to configurations of ring-sizes 12, 13 and 14.

By the end of 1975, their work on the discharging method had run into trouble. Structural changes would be necessary, requiring significant modifications to the program. The problem was that, while trying to disperse the positive charge on each pentagon to its immediate neighbours, they regularly came up against barriers of hexagons, all with zero charge. For Wolfgang Haken, taking a vacation is doing mathematics twenty-three hours a day in a different location, and while he was walking on the beach during his annual family holiday at Key West in Florida he asked himself why the positive charge on the pentagons should not be permitted to 'jump over' these hexagon barriers. This would make for a more efficient process, but would leave Appel and Haken with a dilemma: should they rewrite their program from scratch, or should they adopt a more ad hoc approach? Since the former would have been a horrendous task, they opted for the latter, implementing the final version of the discharging process by hand. This would inevitably require much work, but it would give them extra flexibility by enabling them to make minor changes whenever necessary. As it turned out, this led to so many improvements that they could restrict all their configurations to ring-size 14 or less.

THE FINAL ONSLAUGHT

Throughout the first half of 1976, Appel and Haken worked on the final details of the discharging procedure that would give them their unavoidable set of reducible configurations. To do this, they sought out problem configurations that would necessitate further changes to the discharging procedure. Whenever they found such a configuration, they immediately tested it for reducibility – which could usually be done fairly quickly – and in this way, the reducibility testing by computer could keep pace with the manual construction of the discharging procedure. In the event, the final process used 487 discharging rules, requiring the investigation by hand of about ten thousand neighbourhoods of countries with positive charge, and the reducibility testing by computer of some two thousand configurations.

Reducible configurations seemed to be everywhere, like trees in a forest, but Appel was never quite sure . . .

if you shoot a gun in any direction it's eventually going to hit a tree . . . there's a tree there no matter where you look. But yet you haven't got enough of a description of a forest to formally prove there is a tree there, and the thing you always worry about is there is one little direction where you just shoot it and there is no tree, and it goes out of the forest entirely . . . I just think of that – being in a forest and shooting a gun.

Because the reducibility of an awkward configuration could sometimes take a long time to check, and with memories of the non-reducible Shimamoto horseshoe (which took twenty-six hours), they found it convenient to impose on each configuration an artificial limit of ninety minutes checking time on an IBM 370–158 computer, or thirty minutes on an IBM 370–168 computer. If a configuration could not be proved reducible in this time, it was abandoned and replaced by other configurations: finding such alternative configurations was usually straightforward. By way of comparison, they estimated that to check the computer output of one of the more difficult configurations in full detail would take someone working a forty-hour week about five years in total.

The last few months were indeed extremely heavy on computer time, but here Appel, Haken and Koch were very fortunate. Probably no other institution would have allowed them twelve hundred hours of computer time, especially since the outcome of the whole process was still uncertain. But the University of Illinois computer centre was very supportive throughout, and included the four-colour team among a small group of computer users who were assigned spare computer time when the facility was not being used for anything else. Use was also made of the computer at the university's Chicago campus: programs were sent to Chicago overnight, and the returns were ready by the next morning.

In March 1976, the university's administration unit acquired a powerful new computer. As Haken recalls, local politicians started to ask, 'Why does this administration need a larger computer than the scientists?', and the administrators had to promise, 'OK, half the time, or whenever the thing is not needed by us, the scientists can use it.' Since Appel seemed to be the one scientist who could get the machine to run properly, he had almost exclusive use of it at first, and he benefited from a valuable fifty hours of computer time over the Easter vacation. Everyone was happy: the administrators could claim a more balanced loading of the system over a twenty-four-hour period, while Appel got all the computer time he needed.

In the event, the new computer proved to be so powerful that everything proceeded far more quickly than they had expected, saving Appel and Haken, by their own estimation, a full two years on the reducibility testing. Meanwhile, with the help of Haken's daughter Dorothea, they began an extremely exhausting and stressful few months working in detail through the two thousand or so configurations that would eventually form the unavoidable set:

one of us wrote a section, and then the second one was reading it over, checking it for mistakes, and bringing the errors to the attention of the one who wrote it, and then the third person read it a third time. Reading it over was more strenuous than writing it, because it went faster, and it was more exhausting. It was a terrific amount of work for three people.

Suddenly, by late June, almost before they realized what was happening, the entire job was finished. The Hakens had completed the construction of the unavoidable set, and within two days Appel was able to test the final configuration for reducibility. He celebrated their achievement by placing a notice on the department's blackboard:

> *Modulo careful checking,*
>
> *it appears that*
>
> *four colors suffice*

This phrase, 'four colors suffice', subsequently became the mathematics department's postal meter slogan:

FOUR COLORS SUFFICE

A RACE AGAINST TIME

All that remained was the detailed checking, which had to be done quickly. Not expecting the reducibility testing to advance so well, Appel had arranged a sabbatical visit to Montpellier, where he would be working with Jean Mayer, a professor of French literature with a passion for map colouring. Appel was due to leave in late July. 'We have five weeks', he said. 'Either we announce our result in five weeks, or we have to wait five months.'

Time was indeed of the essence: unknown to them, several other map-colourers were close to solving the problem. At the University of Waterloo in Ontario, Frank Allaire had the best reducibility methods around. These were inherited from Jean Mayer and were, in Haken's words, 'even better than Heesch's and much better than ours'. Appel and Haken were aware of Allaire's superior methods, and on one occasion Appel had written to him for confirmation that an awkward configuration was indeed reducible; this was the only occasion on which they asked for Allaire's help. By 1976, Allaire was several months ahead of them in his investigations into reducibility, and was expecting to complete his solution within a few months.

Meanwhile, at the University of Rhodesia (now Zimbabwe) was a former chemist, Ted Swart, who had carried out the first radio-carbon dating in Africa. He was working independently on the four-colour problem, and was making excellent progress. Swart submitted a paper to the *Journal of Combinatorial Theory* and received a reply from the editor-in-chief, Bill Tutte (also at Waterloo), to the effect that Allaire was working along very similar lines. Allaire and Swart pooled their results, and submitted a paper just before the Appel–Haken proof was announced. This paper described an algorithm for determining the reducibility of configurations, and included a full list of all reducible configurations

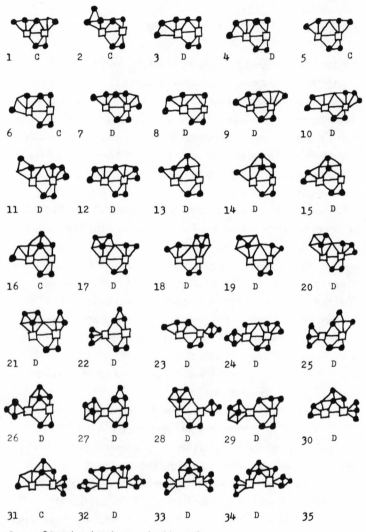

Some of Appel and Haken's reducible configurations.

with ring-size 10 or less. And there were other rivals. The Harvard doctoral student Walter Stromquist had been developing some powerful new methods for tackling the four-colour problem, and was expected to complete his solution within a year, while Frank Bernhart (whom we met briefly in Chapter 8) was also contributing some impressive work on reducibility arguments.

Although no one was aware just how close everyone else was to completing their solutions, Appel and Haken suspected that it would be too risky to wait five months, especially if a rumour were to leak out that they had almost reached their goal. They guessed that Allaire, with his superior reducibility methods, was some way ahead of them on that score, but were confident that their strategy of spending nine-tenths of their time on unavoidable sets and only one-tenth on reducible configurations, while others operated the other way round, gave them an edge.

Even so, they had no time to lose. Drafting in five of their children to help – Dorothea and Armin Haken, and Laurel, Peter and Andrew Appel – they immediately set to work. They had high expectations that, though the checking might turn up many typos and the occasional bad configuration that would need replacing, there would be no major disasters.

Laurel Appel checked through the seven hundred pages of detailed working and found about one error per page, most of them typos: all but fifty of these she immediately corrected herself. Appel then spent the Fourth of July weekend recomputing these fifty cases, and only twelve of them now failed to work. These twelve were replaced by twenty new ones, which were then reduced to just two, and so the process continued. As Haken later remarked:

Somebody has worked one month full-time and found eight hundred mistakes. And then we needed only five days to repair all that. This looks so stable – it is this incredible stability . . .

Appel and Haken knew by this stage that they were safe: even if a few configurations turned out not to be reducible after all, there was more than enough self-correction in the whole system for these configurations to be replaceable easily and quickly. It was not possible for a single faulty configuration, if one existed, to cause the entire edifice to collapse. What is more, since they could adapt their list to produce many hundreds of unavoidable sets of reducible configurations, they had not just one proof, but hundreds of proofs of the four-colour theorem!

By now very confident, they decided to go public. On 22 July 1976, just a few days before Appel was due to leave for France, they formally told their colleagues and sent out complete preprints to everyone in the field who might be approached by other mathematicians or the press for comment. The last thing they wanted was a grumpy 'Well, no one told me' kind of response.

One recipient of a preprint was Bill Tutte. Two years earlier he had written an article in the *American Scientist* claiming that people who used their approach were real optimists since the method had then seemed extremely unlikely to work. But when Tutte heard the news he waxed eloquent, comparing their achievement with the slaying of a fabled Norwegian sea monster:

Wolfgang Haken
Smote the Kraken
One! Two! Three! Four!
Quoth he: 'The monster is no more.'

And when Tutte was interviewed by the press, he told them, 'If they say they've done it, I have no doubt that they've done it.'

Appel and Haken were delighted that a mathematician of Tutte's stature should have given his positive support so quickly. Tutte's endorsement would go a long way to setting peoples' minds at rest, whereas a lukewarm response could have raised serious doubts about their solution.

AFTERMATH

After the blackboard announcement in early July, Andrew Ortoni, a stringer for *The Times* of London who worked at the University of Illinois, asked Appel and Haken's permission to break the news. They asked him to wait until the checking was complete, but promised to let him know as soon as they were ready to announce. On 23 July 1976, the following report duly appeared in *The Times*:

Two American mathematicians have just announced that they have solved a proposition that has been puzzling their kind for more than 100 years . . . Their proof, published today, runs to 100 pages of summary, 100 pages of detail and a further 700 pages of back-up work. It took each of them about 40 hours research a week and 1000 hours of computer time. Their proof contains 10,000 diagrams, and the computer printout stands four feet high on the floor.

Other papers around the world, from the Japanese *Asahi Shinbun* to the German *Neue Zürcher Zeitung*, latched on to the story and there was great excitement. The solution was featured in *Time* and *Scientific American*, and some of the configurations appeared on the front cover of *New Scientist*. Only the *New York Times* seemed to have a policy on not publishing announcements of proofs of the four-colour theorem, 'because they were all false anyway'. In September, according to Appel, Lipman Bers, President of the American Mathematical Society, approached the paper, saying:

'Look, you know, you never have published this thing. It's pretty well accepted.' And they said, 'Well, it's too late for us to publish a news story.' And they said, 'Would you write it?' And so Bers wrote an editorial in the *Times* which acknowledged it and congratulated us.

第3種郵便物認可　　　　　　　　　　　眞

変わる証明の概念

電算機利用は不可欠に

「四色問題」解決の意義

どんな地図でも4色で塗り分けられる

精神衛生上の大問題」となる。ぜなら、一般の数学ファンがこれでもかとれでもかと"証明成功"の論文を数学者たちに持ち込んで点検を迫り、悩まりかけた。

「たとえば、負の三分の一問題、これは定規とコンパスではできない。できないことを数学者が証明するのだから、どんな証明を持ちこまれても、ただちにクズカゴへ捨ててしまう。しかし、四色問題は、そうはいかなかった。大多数の数学者は、四色問題を証明したという数学ファンの論文を、自実のところを持ちつつ、クズカゴに捨てていたのです」と矢野健太郎東工大名誉教授はいう。数学者の悩みの種は、まだかなり残されている。「中の島の両側にかかった七つの橋を、すべて二度づつからず渡る方法はあるか」という有名な「ケーニヒスベルグの七つの橋の問題」など、数学ファンの興味をそそるものだ。

イリノイ大学の二人の数学教授が証明した方法は、数学界に新しい論争を呼び起こしている。

四色問題の証明は、数学の専門家だけではなく、一般の数学ファンにとっても関心の深い問題だった。「最初は大学生だった」といわれている。「平面または球面上のすべての地図は四色で塗り分けられる。これを証明せよ」という問題は、それほど難しい数式を使わなくても取り組みやすそうに見えるが、肯定も否定もせずに何百年も未解決であることは数学上の大前、数学者にこの問題を持ち込ん

で証明をたのんだのは、地図印刷業者が経験的に知っていたことを小耳にはさんだ大学生だった。ちには簡単に解けそうに思えた。が、やってみると難しかった。

このように、ある定理の真偽を問題であるだけでなく「数学者のい論争を呼び起こしている。

紙と鉛筆と人間の理性だけでスマートに解くのではなく、電子計算機を動員して、膨大な計算をした結果はたして、本当の証明が果たして、本当の証明といえるかどうという議論に、一松信京大数理解析研究所教授は、電子計算機の功罪を次のようにいう。

まず、難しかった問題を電算機にやらせて、証明できたとする。証明できたとわかっていれば、数学者は、電子計算機にしいすれはスマートな証明法が見つかるとファイトを燃やせる。「こうししかないような複雑な問題に残っていないと予想されるからだ。

しかし、現在、数学者は電子計算機を十分に使えない状況だ。イリノイ大学の二教授の電算機の成功は、たまたま新鋭の電算機を千二百時間もぶっ続けに使える幸運に恵まれたからだ、という見方もある。

「定理の発見者、証明者、証明年月日のほかに、数学の論文には、もう一つのデータが不可欠となった。その定理を証明するのにいくらかかったかを明記する必要があるだろう」と京大八田カナダるわけです」と佐藤大八郎かすなついている。

カーネギー・メロン大学で開かれたコンピュータ数学のシンポジウムでも、この点が大論議になった。電子計算機で証明するという新しい方法は、将来の数学にとって必要不可欠な方法となる、とみる数学者はふえている。数学者自身「証明」の定義と概念を変えるよう迫られているのだ。なぜならスマートな証明ができるような問題は、ほとんどすべて証明ずみとなり、あとは電算機に頼るしかないような複雑な問題しか残っていないと予想されるからだ。

しかし、反論は電子計算機以外の科学者の使用率が圧倒的に高いうえに、巨額の費用と時間がかかる。電算機の使用を最初から敬遠せざるを得ない数学者も多い。イリノイ大学の二教授の

（西村　幹夫記者）

A Japanese newspaper reports on the Appel–Haken solution.

The American Mathematical Society also published a two-page research announcement by Appel and Haken, in the September 1976 issue of their *Bulletin*, outlining the main ideas of the proof.

Appel and Haken decided to submit the full solution to the *Illinois Journal of Mathematics*. This was not because they felt that they had a better chance of publishing it locally, for if their proof were correct then publication was not in doubt. A few years earlier they had published a long and turgid article in its pages, and they felt that they owed it to the *Illinois Journal* to give it the kudos for publishing their solution of this famous problem. But their main reason for favouring local publication was because they wanted to be in a position to suggest appropriate referees for their proof. It was clearly in everyone's interest – theirs and the *Journal*'s – that a paper such as this should be thoroughly refereed:

We said to the *Illinois Journal*, 'Look, if this thing is wrong, we want to know it as badly as anyone else does, and you're not going to reject it on the grounds of triviality.'

They agreed with the *Journal*'s editor that the best people in the world should be selected as referees – Frank Allaire for the reducibility part and Jean Mayer for the discharging arguments.

In France, Appel spent much of his sabbatical in Montpellier working painstakingly through the details of the proof with Jean Mayer, one step at a time, dealing with all of Mayer's queries and suggestions as they arose. On his return in December 1976, he and Haken set to work refining the details of the proof, conferring with the referees, and preparing the paper and its associated microfiche for publication.

Understandably, Frank Allaire was bitterly disappointed when Appel and Haken announced their result, as he himself was nearing a solution. Moreover, he and Ted Swart had been taking a more systematic approach, as opposed to what they regarded as the hit-and-miss methods of Appel and Haken. But he quickly

suppressed his own feelings for the general good, refereeing the
reducibility part conscientiously and constructively.

Allaire first compared their configurations with his own, finding
a substantial number that appeared on both lists. He next checked
the C-reducible configurations that were not D-reducible. For each
such configuration, one can calculate the number of 'good' colourings of the surrounding ring – those colourings that extend
directly, or after Kempe-chain interchanges of colour, to the configuration inside: for C-reducible configurations of ring-size 14,
about 50000 of the 199291 possible colourings turn out to be
good colourings. Allaire took 400 of these configurations and compared his own numbers with those obtained by Appel and Haken.
In every case the numbers agreed exactly.

Heinrich Heesch was also upset that Appel and Haken had got
there first – not surprisingly, since their method was so closely
based on his, and solving the four-colour problem had been his
goal for more than forty years. But later he became very co-
operative, sending Haken his own list of 2669 reducible configurations, together with supporting data on the number of good
configurations. In September 1977, shortly before publication, a
few discrepancies were noticed, causing Appel and Haken some
alarm. However, Heesch recomputed his configurations and was
soon able to confirm that in every case Appel and Haken were
correct.

Appel and Haken's published paper was a substantial improvement on the rough-and-ready preprint they had sent out in July
1976. In particular, they discovered that their preprint contained
several 'repeats' of configurations. They also found many
instances of one configuration appearing inside another, so that in
each case they could omit the larger one. By weeding out these
superfluous configurations they managed to reduce the original
list of 1936 reducible configurations to the 1482 configurations of
the published version; they later reduced this number still further

to 1405. However, they saw no point in aiming for the 'smallest possible' number if this greatly increased the total computer time. As Appel pointed out,

If one configuration replaces twelve, but that one configuration takes two hours and the twelve take five minutes, that doesn't make sense.

Appel and Haken's solution appeared in two parts, in the December 1977 issue of the *Illinois Journal of Mathematics*. Part I was 'Discharging', written by the two of them: it outlined the overall strategy of their proof and described their methods of discharging for constructing the unavoidable set. Part II, 'Reducibility', was written with John Koch: it described the computer implementation and listed the entire unavoidable set of reducible configurations. The two papers were supplemented by a microfiche containing 450 pages of further diagrams and detailed explanations.

Appel and Haken had achieved their goal: the four-colour theorem was proved.

... *but is it a proof?*

The Appel–Haken proof of the four-colour theorem was greeted with great enthusiasm: after 124 years, one of the most famous problems in mathematics had finally been solved. It was also greeted with scepticism:

we've been here before so many times – remember the Shimamoto horseshoe.

It was greeted with outright rejection:

in my view such a solution does not belong to mathematical science at all.

But more than anything else, it was greeted with great dis-appointment:

God wouldn't let the theorem be proved by a method as terrible as that!

Too late – he already had!

COOL REACTION

In August 1976 the American Mathematical Society and the Mathematical Association of America held a joint summer meeting at the University of Toronto, and Wolfgang Haken was one of the speakers. In a report on the event, Donald Albers described the scene:

The elegant and old lecture hall was jammed with mathematicians anxious to hear Professor Haken give the proof. It seemed like the perfect setting to announce a great mathematical result. He proceeded to outline clearly the computer-assisted proof that he and his colleagues had devised. At the conclusion of his remarks, I had expected the audience to erupt with a great ovation. Instead, they responded with polite applause!

Reaction to Haken's lecture was certainly mixed. Ted Swart, who was visiting Waterloo at the time, attended the meeting and praised the quality of Haken's presentation, recalling that the applause was a little more than merely polite. Attending with him were Frank Allaire, Frank Bernhart and Bill Tutte. As Swart recalled:

After it was all over, the two Franks and I huddled together and agreed among ourselves that Haken *et al.* 'had done it'.

. . . and many people asked them about the proof:

What did we think? Did we think they had really done it? And I can only tell you that we were unanimous in our view that they'd done it. Unanimous.

But this was by no means the general reaction. According to Albers:

Mathematician after mathematician expressed uneasiness with a proof in which a computer played a major role. They were bothered by the fact that more than 1000 hours of computer time had been expended in checking some 100,000 cases and often suggested (hoped?) that there might be an error buried in the hundreds of pages of computer printouts. Beyond that concern was a hope that a much shorter proof could be found.

Indeed, acceptance of the computer part of the proof seemed to be based partly on age. Haken's son Armin, by then a graduate student at the University of California at Berkeley, gave a lecture on the four-colour problem and on his contribution to its solution. At the end, the audience split into two groups: the over-forties could not be convinced that a proof by computer was correct, while the under-forties could not be convinced that a proof containing 700 pages of hand-calculations could be correct.

It was clear that Appel and Haken's solution of the four-colour problem had created something new in mathematics. Is it a proof? And if so, how do we know that it is a proof? Albers concluded,

it seems that the computer-assisted work of Appel, Haken and Koch on the well-known Four-Color Problem may represent a watershed in the history of mathematics. Their work has been remarkably successful in forcing us to ask What is a Proof Today?

WHAT IS A PROOF TODAY?

The concerns about the Appel–Haken proof, and particularly about the use of the computer, continued to rumble on – and still do. Can a proof be considered valid if it cannot be checked by hand? In February 1979 a philosopher, Thomas Tymoczko, set the cat among the pigeons when he wrote a paper for *The Journal of Philosophy* entitled 'The four-color problem and its philosophical significance'. In this paper, and in others written around the same time, he asked whether the proof of the four-colour theorem could be regarded as valid, relying as it did on extensive use of a computer. He fully accepted that Appel and Haken had *demonstrated* that four colours suffice for all maps, but felt that it was not a *proof* in the sense in which mathematical proofs were usually understood.

Among the criteria Tymoczko laid down for a valid proof were that it should be both *convincing* and *surveyable*. With the former he had no quarrel:

most mathematicians have accepted the 4CT, and none, to my knowledge, has argued against it.

It was with the latter criterion that he mainly took issue, asserting with full force and in great detail that the Appel–Haken proof is not surveyable – that is, it cannot be checked in all its detail:

No mathematician has seen a proof of the 4CT, nor has any seen a proof that it has a proof. Moreover, it is very unlikely that any mathematician will ever see a proof of the 4CT.

Tymoczko had no problem with the discharging procedure for producing an unavoidable set. This had been applied by hand to yield the unavoidable set of configurations, and Appel and Haken had given a rigorous and surveyable proof that it did indeed yield an unavoidable set. On the other hand, no mathematician could survey the reducibility part of the proof, because the details of the method of proof are hidden inside the computer. No appeal to computers can be considered a valid method of proof – in particular, computers can make mistakes. They might malfunction or be misprogrammed, their output might be misread, or the programs might not capture the mathematical intention – the Shimamoto episode is an example of where the computer got it wrong. If such 'experiments' with computers are allowed, mathematics is in danger of becoming an empirical science, as fallible as physics. Tymoczko concluded that if we accept Appel and Haken's proof as valid, then we are forced to modify our concept of 'proof' to allow computer experimentation as a new method for establishing a mathematical result.

Tymoczko's article caused a storm of protest, both within the columns of *The Journal of Philosophy* and elsewhere. One person who was particularly riled by his views was Ted Swart, who believed that Tymoczko had it all wrong. Within a matter of days he had written a lengthy response, which he then submitted to *The Journal of Philosophy*, but without success:

Whether for reasons of professional jealousy or whatever, the editor did me the discourtesy of rejecting it outright without even submitting it to referees. Fortunately I had sent a copy to Martin Gardner who responded right away and said it would be a tragedy if my excellent article was not published. He went on to suggest that I should submit it to the *American Mathematical Monthly*, indicating that he had recommended I should do so.

The *Monthly* had the article in print within weeks, and a few months later Ted Swart was awarded a Lester R. Ford Award by the Mathematical Association of America for outstanding expository writing.

In his article, Swart first pointed out that all those who worked on reducibility testing were happy with Appel, Haken and Koch's reducibility results, because they had to a large extent been checked independently by the use of different reducibility testing programs on different computers. On the general question of using a computer, he later commented that:

For the most part I regard computer-assisted proof as just an extension of pencil and paper. I don't think there's some great divide which says that OK, you're allowed to use pencil and paper but you're not allowed to use a computer because that changes the character of the proof. I don't see that myself. I find such an argument strange.

Indeed, if Swart had a worry about the Appel–Haken proof, it lay in the discharging part, where a computer had *not* been used:

It is no easy matter to check that the final working list of 1482 unavoidable configurations embraces all the unavoidable configurations arising from their discharging procedure. Thus the Haken and Appel proof is subject to some degree of uncertainty, and Tymoczko is definitely incorrect when he says that no mathematician 'has argued against it'.

As mathematical proofs become longer and computers are used more widely, the whole question of surveyability has become fraught with difficulties. A computer that performs extensive but completely routine tasks may certainly be considered at least as reliable as a human who checks by hand a proof that is long and complicated or breaks down into a large number of special cases. In Ted Swart's words, 'Human beings get tired, and their attention wanders, and they are all too prone to slips of various kinds . . . Computers do not get tired.' To illustrate the contrast, let us look at two celebrated proofs from an area of mathematics called group theory.

In 1963 Walter Feit and John Thompson proved a remarkable theorem in group theory. Its published proof runs to over 250 pages of concentrated argument, and initially contained a number of minor errors that were subsequently corrected. Although the possibility of human error in such a lengthy proof is substantial, their proof is generally accepted by mathematicians, most of whom have never worked through it in detail: they are happy to trust all those who *have* checked it thoroughly. Much the same can be said for Andrew Wiles's recent proof of Fermat's last theorem, a long-standing problem in number theory. While it is possible that either proof may subsequently be found to be in error (as was Kempe's proof after eleven years), each is currently regarded as being correct.

The problem of classifying all finite simple groups, a major area of study from 1950 to 1980, provides a contrast. It was proved that these objects called finite simple groups can be classified into eighteen general types, together with twenty-six separate examples. The work amounted to many thousands of pages by hundreds of contributors, and parts of it made substantial use of computers. One of the leaders in this classification was Daniel Gorenstein, who wrote in 1979 that:

This is an appropriate moment to add a cautionary word about the meaning of 'proof' in the present context; for it seems beyond human capacity to present a clearly-reasoned several-hundred-page argument with absolute accuracy . . . there are no guarantees – one must live with this reality.

In spite of this caution, most mathematicians have been happy to accept the classification as fully proved.

But surveyability has not been the only question hanging over the Appel–Haken proof. Several mathematicians have complained that the proof is not sufficiently transparent. Among them was Ian Stewart, the well-known writer and broadcaster on mathematics, who complained that it did not explain *why* the theorem is true – partly because it was too long for anyone to grasp all the details, and partly because it seemed to have no structure:

The answer appears as a kind of monstrous coincidence. Why is there an unavoidable set of reducible configurations? The best answer at the present time is: there just is. The proof: here it is, see for yourself. The mathematician's search for hidden structure, his pattern-binding urge, is frustrated.

Even more forceful was Daniel Cohen, who had himself worked on the four-colour problem:

In the analysis of each case the program only announced whether or not the procedure terminated successfully. The entire output from the machine was a sequence of yeses. This must be distinguished from a program which produces a quantity as output which can subsequently be verified by humans as being the correct answer . . . The real thrill of mathematics is to show that as a feat of pure reasoning it can be understood why four colors suffice. Admitting the computer shenanigans of Appel and Haken to the ranks of mathematics would only leave us intellectually unfulfilled.

H. S. M. Coxeter accepted the inevitable:

It has always seemed to me a different kind of theorem from other kinds of theorems. When the computer proof has been checked by quite a number of people, and they're all satisfied with it, well then we'll have to accept it. But I think it's very unlikely that anyone can break that proof down into something that one would regard as an ordinary proof. So it's rather in a different category than all other theorems.

George Spencer-Brown, on the other hand, refused to believe that Appel and Haken had proved anything:

Nowhere in their long and often irrelevant account do they provide the evidence that would enable the reader to check what they say. It may, or may not, be 'possible' to prove the color theorem the way they claim. What is now certain is that they did not do so . . . not only is no proof to be found in what they published, but there is not anything that even begins to look like a proof. It is the most ridiculous case of 'The King's New Clothes' that has ever disgraced the history of mathematics . . .

What everyone did agree on, however, was that the Appel–Haken proof could not be described as beautiful or elegant mathematics. In his celebrated book *A Mathematician's Apology*, the English mathematician G. H. Hardy had claimed that 'there is no permanent place in the world for ugly mathematics', and even Ken Appel did not demur from this in respect of his own proof:

there were people who said, 'This is terrible mathematics, because mathematics should be clean and elegant', and I would agree. It would be nicer to have clean and elegant proofs.

While everyone would like to see a non-computer solution of the four-colour theorem, such a solution along the lines of the Appel–Haken proof is almost certainly unattainable – not least because E. F. Moore's example in Chapter 9 shows that any unavoidable set of reducible configurations must include configurations with ring-size 12 or more. For a non-computer proof, new ideas are needed, and such ideas have not been forthcoming.

MEANWHILE . . .

After the Toronto meeting, Wolfgang Haken embarked on a lecture tour of the United States, explaining the details of the solution. Ken Appel was doing a similar thing in Europe, giving lectures in France where he was spending time with Jean Mayer, and four lectures at the University of Bristol, where he had spent a sabbatical year some time earlier. These lectures were informative and convincing to those who attended, but others in England took a very different view.

Just before Christmas, Spencer-Brown gave a lecture at London University's Institute of Education in which he cast doubt on the Appel–Haken solution, pointing out, with some justification, that it could not yet be regarded as a proof since its details had not been published. Spencer-Brown had himself worked on the four-colour problem up to 1964, the year of his *Laws of Form*, a book on logic that his friend Bertrand Russell mentioned in his autobiography and which has been variously described as 'a work of genius' and 'pretentious triviality'. The appearance of Appel and Haken's proof spurred him to re-enter the fray, and he worked night and day towards a solution that did not rely on a computer.

His lecture seemed, to me at least, a bizarre occasion. An entry fee was charged, and media people turned up in force and drank champagne at the front of the lecture theatre while the rest of the audience sat and waited. Spencer-Brown's proposed solution involved Kempe-chain arguments with an ingenious new twist: the colours chosen were *red, blue, red-and-blue (purple)* and *neither-red-nor-blue (white)*, where the colour *purple* could play the role of either *red* or *blue* whenever two Kempe-chains met. A video was made of the lecture: this recording was intended to be 'the first published proof of the theorem', but in the event so many details were omitted that it could not be counted as such. Following this event, Spencer-Brown was invited to Stanford University for several months, where his proof was scrutinized, found to be deficient, and repeatedly corrected and re-scrutinized. The current version of his solution can be found (in English) as an appendix to the German edition of his *Laws of Form*.

Following the appearance of their unorthodox proof, Appel and Haken were sometimes made to feel most unwelcome. The most dramatic instance was when the head of a mathematics department refused to allow them to meet his graduate students, on the grounds that:

Since the problem had been taken care of by a totally inappropriate means, no first-rate mathematician would now work any more on it, because he would not be the first one to do it, and therefore a decent proof might be delayed indefinitely. It would certainly require a first-rate mathematician to find a satisfactory proof, and that was now impossible.

They were made to feel that they had done something very wicked, and that the innocent souls of students needed to be protected from their bad influence.

That head of department may have had a point. In 1977 Frank Allaire had attended the Seventh Manitoba Conference on Numerical Mathematics and Computing, and presented his proof of the four-colour theorem. Because of his different discharging method and his superior methods for reducibility, his solution had required only fifty hours of computer time. But although he produced a lengthy paper for the conference proceedings, containing many of the main ideas of his proof, nothing ever appeared in a refereed journal: had it appeared, it could have provided independent corroboration for the Appel–Haken proof.

Two other publications appeared in 1977. For the readers of *Scientific American*, Appel and Haken wrote 'The solution of the four-color-map problem', still one of the clearest descriptions of their approach. Around the same time, a second book on the four-colour problem was published. *The Four-Color Problem: Assaults and Conquests*, by Thomas Saaty and Paul Kainen, was based on an award-winning paper by Saaty that had appeared in the *American Mathematical Monthly*. However, the book's passage to publication was interrupted, as its preface records:

A funny thing happened to this book on the way to the publisher. More than two years ago, we had completed the manuscript of a book entitled *Assaults on the Four-Color Conjecture* . . . Our original book was oriented towards the variety of approaches with which the four-color conjecture (4CC for brevity) has been investigated. The reader can well imagine our mixed reactions to the announcement, in the summer of 1976, that the 4CC had been proved – by a computer, no less.

Saaty and Kainen retrieved their manuscript and revised it in the light of Appel and Haken's solution. When their book eventually appeared, it became a best-seller.

By the early 1980s, rumours were beginning to spread that there was a major error in Appel and Haken's proof of the four-colour theorem. In view of its complexity, many errors might have been expected to appear, but this did not happen. The only significant one, which probably caused the rumours, was discovered in 1981 by Ulrich Schmidt, an electrical engineering student in Aachen. The four-colour problem interested him because there were parallels with checking the design of computer chips. For his Diplomarbeit (master's thesis) at the Aachen Technische Hochschule, Schmidt spent one year (the maximum permitted time) working through 40 per cent of the discharging part of the Appel–Haken proof. He found one error, which took Haken about two weeks to correct, in addition to a few misprints. In 1985 a further error – a minor drawing error in one configuration – was found by S. Saeki in Japan. Other than these, and a few more misprints, no significant error has ever been found in the Appel–Haken proof.

In 1986, Appel and Haken received a letter from the editor of the *Mathematical Intelligencer*, who had heard the persistent rumours that something was wrong with the proof, and invited them to set the record straight. Welcoming such an opportunity, they replied:

With this kind offer we cannot but comply, although it is a pity to spike a good rumor. Mathematical rumors do add interest and excitement to the conversation at mathematical meetings. However, the rumors about the Four Color Theorem seem to be based on a

misinterpretation of the results of the independent check of details
of the proof by U. Schmidt.

The result was 'The four color proof suffices', an upbeat article
describing their methods in some detail and discussing how they
had corrected the error that Schmidt had identified. In 1989 there
appeared their last word on the subject, a hefty tome entitled *Every
Planar Map is Four Colorable*, which supplied several more details
of the proof, proved some related results, corrected all the errors
that had been discovered, and included a printed version of all
their microfiche pages.

A NEW PROOF

In 1994 Neil Robertson, Daniel Sanders, Paul Seymour and Robin
Thomas added an exciting new chapter to the four-colour saga.
For the previous ten years, they had come up with some spectacular results in graph theory, and they now turned to the four-colour
problem because it was related to other things they were working
on.

They were concerned that Appel and Haken's proof was still not
fully accepted, and believed that there was still doubt about its validity. The main reason was not that the proof required a computer,
even though a full check of the reducibility part would have
involved inputting 1482 configurations by hand into the computer
and a lot of programming to test each one for reducibility. It was
the unavoidability arguments that caused them most concern: the
discharging algorithm that Appel and Haken had carried out by
hand was very complicated, and no one had ever made an independent check of it all.

After a week of working through the details of the Appel–Haken
proof, they gave up. They decided that it would be much more fun,

and more instructive, to pursue a proof of their own, using the same general approach that Appel and Haken had taken. It took them a whole year to complete the work, and the resulting proof was shorter and more robust than its predecessor.

The unavoidable set that they obtained contains only 633 reducible configurations – they could have reduced it further, to just 591 configurations, but only at the expense of more computer time. Moreover, their discharging procedure for proving unavoidability required only 32 discharging rules, in contrast to the 487 rules used by Appel and Haken. Because they considered computer proofs of long and complicated results to be *more* reliable than hand-checked ones, they used the computer for both the unavoidability and reducibility parts of the proof. (What would Tymoczko have said!) What is more, all the steps in their proof can be externally verified by anyone on their home computer in about three hours.

THE FUTURE . . .

Now that the four-colour problem has been solved, what else is there for map-colourers to do? Bill Tutte asked this question back in 1978:

I imagine one of them outgribing in despair, crying 'What shall I do now?' To which the proper answer is 'Be of good cheer. You can continue in the same general line of research.'

For the four-colour theorem is by no means the end of the line – in fact, it is more of a beginning, as there are several mathematical problems that extend the four-colour theorem and develop its ideas in new and exciting directions. In spite of its enormous difficulty, the four-colour problem is just one special instance of some much harder problems, and on these problems good progress is already being made.

With these thoughts of the future in our minds, we leave our last poetic musings to Bill Tutte:

The Four Colour Theorem is the tip of the iceberg,
the thin end of the wedge
and the first cuckoo of Spring.

Notes and references

FURTHER READING

For an overview of the history of the four-colour problem up to 1936, including reprints of several of the papers described here, see the 1998 paperback edition of N. L. Biggs, E. K. Lloyd and R. J. Wilson, *Graph Theory 1736–1936*, Clarendon Press, Oxford. This book is referred to below as *BLW*.

A detailed account of the underlying mathematics in the language of graph theory is given in Rudolf Fritsch and Gerda Fritsch, *The Four-Color Theorem: History, Topological Foundations, and Ideas of Proof*, Springer, 1999; Chapter 1 contains biographies of several of the mathematicians featured here. This book is referred to below as *FF*.

A well-written recent textbook on graph theory, concentrating particularly on the four-colour problem, is by my namesake, Robert A. Wilson, *Graphs, Colourings and the Four-Colour Theorem*, Oxford Science Publications, 2002.

Earlier influential books on the four-colour problem are O. Ore, *The Four-Color Problem*, Academic Press, 1967, and Thomas L. Saaty and Paul C. Kainen, *The Four-Color Problem: Assaults and Conquest*, McGraw-Hill, 1977 (reprinted in paperback by Dover, 1986).

For Kenneth Appel and Wolfgang Haken's own accounts of the history of the four-colour problem and of their part in its solution, see 'The solution of the four-color-map problem', *Scientific American* **237** No. 4

(October 1977), 108–21, and 'The four-color problem', in *Mathematics Today* (ed. L. A. Steen), Springer (1978), 153–80.

There is also an excellent account of the history and proof of the four-colour problem, especially on the work of Heesch, Haken and Appel and the philosophical aspects of the solution, in Donald MacKenzie, 'Slaying the Kraken: the sociohistory of a mathematical proof', *Social Studies of Science* **29** (1) (February 1999), 7–60.

Another invaluable source for the writing of this book has been the unpublished manuscript of a proposed book, *Origins of Graph Theory*, by P. J. Federico, who unfortunately died before he was able to bring his project to completion.

Biographies of several of the mathematicians mentioned in this book appear in the *Dictionary of Scientific Biography* (ed. C. C. Gillespie), Scribner's, New York, 1970–1990 (referred to below as *DSB*), and in the four-volume version, *Biographical Dictionary of Mathematicians* (1991), extracted from the *DSB*. Among the many general histories of mathematics that can be recommended are Dirk J. Struik, *A Concise History of Mathematics* (4th edn), Dover Publications, 1967, and Victor J. Katz, *A History of Mathematics: An Introduction* (2nd edn), Addison-Wesley, 1998.

DETAILED NOTES AND REFERENCES

The four-colour problem

2. Kenneth O. May's article on the four-colour problem is 'The origin of the four-color conjecture', *Isis* **56** (1965), 346–8. Another article on the problem's origins is H. S. M. Coxeter's 'The four-color map problem, 1840–1940', *Mathematics Teacher* **52** (1959), 283–9.

3. The book that asserts that graph theory developed primarily from work on the four-colour theorem is M. Aigner, *Graphentheorie – Eine Entwicklung aus dem 4-Farbenproblem*, B. G. Teubner, Stuttgart, 1984.

14. The April Fool's column and the follow-up appeared in Martin Gardner, 'Six sensational discoveries that somehow have escaped public attention', *Scientific American* **232** No. 4 (April 1975), 126–30, and **233** No. 1 (July 1975), 115.

The problem is posed

Biographies of Augustus De Morgan, Sir William Rowan Hamilton, Charles Sanders Peirce and August Ferdinand Möbius appear in the *DSB*, and a biography of Francis Guthrie appears in the *Dictionary of South African Biography* **2** (1972), 279–80. Brief biographies of Francis and Frederick Guthrie, De Morgan, Hamilton, Whewell, Ellis and Peirce appear in Chapter 1 of *FF*.

17. De Morgan's letter to Sir William Rowan Hamilton, sent from 7 Camden Street, Camden Town, on 23 October 1852, is now housed in the archive collection of the library of Trinity College, Dublin (TCD MS 1493/668).

19. Frederick Guthrie's 'Note on the colouring of maps' appeared in the *Proceedings of the Royal Society of Edinburgh* **10** (1880), 727–8.

20. Heinrich Tietze's example appeared in his *Famous Problems of Mathematics*, Graylock Press, New York (1965), 77–8.

21. The meeting between De Morgan and Hamilton is recalled by De Morgan's widow Sophia in S. E. De Morgan, *Memoir of Augustus De Morgan, with Selections of His Writings*, Longman, Green & Co., London (1882), 333.

23. Hamilton's reply to De Morgan was sent from Dunsink Observatory, Dublin, on 26 October 1852.

24. De Morgan's letters to William Whewell and Robert Ellis, dated 9 December 1853 and 24 June 1854, are housed in the Wren Library, Trinity College, Cambridge (Whewell Add. Mss., a.202^{125} and c.67^{111}). A useful discussion of De Morgan's ideas appears in N. L. Biggs, 'De Morgan on map colouring and the separation axiom', *Archive for History of Exact Sciences* **28** (1983), 165–70.

25. De Morgan's anonymous review of W. Whewell's *The Philosophy of Discovery* appears in *The Athenaeum*, No. 1694 (14 April 1860), 501–3. The later rediscovery of this review was by John Wilson, 'New light on the origin of the four-colour conjecture', *Historia Mathematica* **3** (1976), 329. The Hotspur quotation appears in William Shakespeare's *King Henry IV, Part 1*, Act III, Scene I.

26. De Morgan's 3 March 1860 letter to Whewell appears on p. 302 of S. E. De Morgan's book (see p. 21 above).

Peirce's interest in the four-colour problem is described in Chapter 19,

'The four-color problem', in *Studies in the Scientific and Mathematical Philosophy of Charles S. Peirce: Essays by Carolyn Eisele* (ed. R. M. Martin), Mouton, The Hague, 1971, 216–22, and is referred to in Norman L. Biggs, E. Keith Lloyd and Robin J. Wilson, 'C. S. Peirce and De Morgan on the four-colour conjecture', *Historia Mathematica* **4** (1977), 215–16. The Peirce quotation appears on p. 219 of Eisele (see above).

28 The discovery of Listing and Möbius's one-sided surface is outlined in P. Stäckel, 'Die Entdeckung der einseitigen Flächen', *Mathematische Annalen*

30. The origin of Möbius's problem of the five princes is presented in R. Baltzer, 'Eine Erinnering an Möbius und seinen Freund Weiske', *Berichte der Sächsischen Gesellschaft der Wissenschaften zu Leipzig* **37** (1885), 1–6.

(1899), 598–600.

31. The problem of the five palaces is discussed in Tietze's book (see p. 20 above).

36. The content of Baltzer's lecture is summarized in his article (see p. 30 above).

37. Baltzer's error was compounded by F. Dingeldey in *Topologische Studien*, B. G. Teubner, Leipzig, 1890; by Isabel Maddison, 'Note on the history of the map-coloring problem', *Bulletin of the American Mathematical Society* **3** (1896–7), 257; by W. W. Rouse Ball, *Mathematical Recreations and Essays* (11th edn), Macmillan, London (1939), p. 223; and by E. T. Bell, *The Development of Mathematics* (2nd edn), McGraw-Hill (1945), p. 606. The record on Möbius's contribution to the four-colour problem was set straight by H. S. M. Coxeter (see p. 2 above).

Euler's famous formula

Biographies of Leonhard Euler, Augustin-Louis Cauchy and Simon-Antoine-Jean Lhuilier appear in the *DSB*. For a readable account of Euler's life, see William Dunham's *Euler: The Master of Us All*, Mathematical Association of America, 1999.

41. For an excellent account of the history and properties of polyhedra, see Peter R. Cromwell, *Polyhedra*, Cambridge University Press, 1997 (paperback edn, 1999).

For a geometrical proof that there are only five regular polyhedra, see Book XIII in T. L. Heath's *The Thirteen Books of Euclid's Elements*,

Cambridge University Press, 1908. Further details can be found in Cromwell's book (see above).

45. The extract from Euler's letter to Christian Goldbach appeared in P.-H. Fuss, *Correspondance mathématique et physique de quelques célèbres géomètres du XVIIIème siècle*, St Petersburg, 1843; an English translation appears in *BLW*, pp. 76–7. The letter appears in full in A. P. Juškević and E. Winter, *Leonhard Euler und Christian Goldbach: Briefwechsel 1729–1764*, Akademie-Verlag, Berlin, 1965.

46. For accounts of René Descartes's work on polyhedra, see Cromwell's book (see p. 41 above), C. Adam and P. Tannery, *Oeuvres de Descartes*, Vol. 10, Cerf, Paris (1897–1913), 257–77; and P. J. Federico's *Descartes on Polyhedra: A Study of the De Solidorum Elementis*, Springer-Verlag, 1982. Euler's papers on polyhedra are 'Elementa doctrinae solidorum' and 'Demonstratio nonnullarum insignium proprietatum quibus solida hedris planis inclusa sunt praedita', *Novi Commentarii Academiae Scientiarum Imperialis Petropolitanae* **4** (1752–3, publ. 1758), 109–40, 140–60; both papers appear in Vol. 26 of *Leonhardi Euleri Opera Omnia (1)* (ed. A. Speiser), Commentationes Geometricae, Zürich, 1953.

47. Legendre's proof appeared in his *Eléments de géométrie* (1st edn), Firmin Didot, Paris, 1794.

48. Cauchy's paper on polyhedra was 'Recherches sur les polyèdres – premier mémoire', *Journal de l'Ecole Polytechnique* **9** (Cah. 16) (1813), 68–86; an English translation of part of this paper appears in *BLW*, pp. 81–3.

52. The examples given by Lhuilier (whose name is also written as 'L'Huilier') appeared in 'Mémoire sur la polyédrométrie', *Annales de mathématiques* **3** (1812–13), 169–89; an English translation of part of this paper appears in *BLW*, pp. 84–6.

53. The 'only five neighbours theorem' for maps was obtained by Kempe (see Chapter 5), who also derived a result similar to the counting formula.

Cayley revives the problem . . .

Biographies of Arthur Cayley appear in the *DSB* and Chapter 1 of *FF*.

62. Cayley's query appeared in the *Proceedings of the London Mathematical Society* **9** (1877–8), 148; and in *Nature* **18** (1878), 294. His paper 'On the colouring of maps' appears in the *Proceedings of the Royal Geographical Society* **1** (1879), 259–61; it is reprinted in full in *BLW*, pp. 93–4.

65. Levi ben Gerson's use of mathematical induction is discussed in Victor Katz, 'Combinatorics and induction in medieval Hebrew and Islamic mathematics', *Vita Mathematica* (ed. R. Calinger), Mathematical Association of America (1996), 99–106.

5 . . . and Kempe solves it

Biographies of James Joseph Sylvester appear in the *DSB* and the book by Karen Parshall (see p. 75 below), and a biography of Alfred Bray Kempe (by Archibald Geikie) appeared in the *Proceedings of the Royal Society* **102** (1923), i–x. Brief biographies of Sylvester, Kempe and William Story are given in Chapter 1 of *FF*.

75. Sylvester's correspondence with Gilman and others appears in Karen Parshall, *James Joseph Sylvester: Life and Work in Letters*, Clarendon Press, Oxford, 1998.

76. Kempe's first paper was 'On the solution of equations by mechanical means', *Messenger of Mathematics* **2** (1873), 51–2. His 'How to draw a straight line' was first published in four parts in *Nature* **16** (1877), 65–7, 86–9, 125–7 and 145–6.

77. Kempe's paper 'On the geographical problem of the four colours' appeared in the *American Journal of Mathematics* **2** (part 3) (1879), 193–200; it is reprinted in full in *BLW*, pp. 96–102. A preview of it appeared in *Nature* **20** (17 July 1879), 275; and simplified versions appeared in the *Proceedings of the London Mathematical Society* **10** (1878–9), 229–31, and in *Nature* **21** (26 February 1880), 399–400.

90. The term *graph* was first used in this sense by Sylvester in 'Chemistry and algebra', *Nature* **17** (1877–8), 284; this note is reprinted in full in *BLW*, pp. 65–6.

91. Story's 'Note on the preceding paper' appeared in the *American Journal of Mathematics* **2** (1879), 201–4.

92. Sylvester's letter to D. Gilman appears on pp. 195–6 of Parshall's book (see p. 75 above).

Reports of the 5 November and 3 December 1879 meetings of the Johns Hopkins Scientific Association appeared in the *Johns Hopkins University Circular* **1**, No. 2 (January 1880), 16.

93. Peirce's reformulation of the four-colour problem appears on p. 218 of

Eisele (see p. 26 above). The 'Notice by Mr. Peirce' appeared in *The Nation* No. 756 (25 December 1879), 440.

Peirce's New York lecture was mentioned in the *Report of the National Academy of Sciences for the Year 1899* (1900), 12–13.

94. The letter from Story to Peirce appears on p. 359 of Eisele (see p. 26 above).

A chapter of accidents

Biographies of Charles Dodgson, Peter Guthrie Tait, Thomas Penyngton Kirkman and Sir William Rowan Hamilton appear in the *DSB*. For full biographies of Tait and Hamilton, see C. C. Knott, *Life and Scientific Work of Peter Guthrie Tait*, Cambridge University Press, 1911, and R. P. Graves, *Life of Sir William Rowan Hamilton*, Longman, London and Hodges, Figgis & Co., Dublin, 1889. Kirkman's life and works are chronicled in N. L. Biggs, 'T. P. Kirkman, mathematician', *Bulletin of the London Mathematical Society* **13** (1981), 97–120. Brief biographies of Tait and Hamilton appear in Chapter 1 of *FF*.

95. Dodgson's *An Elementary Theory of Determinants* was published by Macmillan in 1867. His puzzle appeared in Stuart D. Collingwood's *The Life and Letters of Lewis Carroll (Rev. C. L. Dodgson)*, Nelson, 1898, 371.

96. Edouard Lucas's translation of Kempe's paper, 'Le Problème géographique des quatre couleurs', appeared in *Revue scientifique (3)* **6** (1883), 12–17.

Lucas's extended discussion of the four-colour problem appeared in *Récréations mathématiques*, Vol. 4, Gauthier-Villars, 1894.

For information about Baltzer's lecture to the Leipzig Scientific Society, see p. 30 above.

97. J. M. Wilson's letters of 1 January 1887 and 1 June 1889 appeared in the *Journal of Education* **9** (1887), 11–12, and **11** (1889), 277. His account of Frederick Temple's solution also appears in *Memoirs of Archbishop Temple* Vol. 1 (ed. E. G. Sandford), Macmillan, London, 1906, 59–60.

98. John Cook Wilson's paper 'On a supposed solution of the "four-colour problem"' appeared in the *Mathematical Gazette* **3** (1904–6), 338–40.

99. The golf ball anecdote is discussed by Chris Denley and Chris

Pritchard in 'The golf ball aerodynamics of Peter Guthrie Tait', *Mathematical Gazette* **78** (1994), 298–313.

For Kempe's note in *Nature*, see p. 77 above.

P. G. Tait's first attempted solutions appeared in 'On the colouring of maps', *Proceedings of the Royal Society of Edinburgh* **10** (1878–80), 501–3.

102. Tait's second note on the subject was 'Remarks on the previous communication' [by Guthrie], *Proceedings of the Royal Society of Edinburgh* **10** (1878–80), 729; it appears in full in *BLW*, p. 104. His letters to Kempe are in the West Sussex Public Record Office, Chichester.

104. For details of Frederick Guthrie's note, see p. 19 above.

105. Tait's paper was 'Note on a theorem in the geometry of position', *Transactions of the Royal Society of Edinburgh* **29** (1880), 657–60.

108. Kirkman's paper 'On the representation of polyhedra' appeared in the *Philosophical Transactions of the Royal Society* **146** (1856), 413–18; part of this paper appears in *BLW*, pp. 29–30.

109. W. R. Hamilton's icosian calculus was described in 'Memorandum representing a new system of roots of unity', *Philosophical Magazine (4)* **12** (1856), 446, and in the *Proceedings of the Royal Irish Academy* **6** (1853–7), 415–16. A summary of the main ideas, together with instructions for playing the icosian game, can be found in *BLW*, pp. 32–5.

112. For details of Tait's paper, see p. 105 above.

Tait's final paper on the subject was 'Listing's *Topologie*', *Philosophical Magazine (5)* **17** (1884), 30–46.

113. Kirkman's poem and discussion appeared as Question 6610 with Solution by the proposer in *Mathematical Questions and Solutions from the Educational Times* **35** (1881), 112–16.

114. W. T. Tutte's cubic polyhedron appeared in 'On Hamiltonian circuits', *Journal of the London Mathematical Society* **21** (1946), 98–101.

115. The poem about Bill Tutte is by Norman Biggs.

7 *A bombshell from Durham*

A biography of Julius Petersen appears in the *DSB*. A full account of Heawood's life is G. A. Dirac's 'Percy John Heawood', *Journal of the London Mathematical Society* **38** (1963), 263–77; a brief obituary by J. Duff appeared in *Nature* **175** (1955), 368, and a brief biography appears in Chapter 1 of *FF*.

119. Heawood's letter to Alfred Errera appeared in Errera's 'Exposé historique du problème des quatre couleurs', *Periodica Mathematica (4)* **7** (1927), 20–41.

Heawood's 'Map-colour theorem' appeared in the *Quarterly Journal of Pure and Applied Mathematics* **24** (1890), 332–8; most of it can be found in *BLW*, pp. 105–7, 112–15.

123. Kempe's admission of his error is reported in the *Proceedings of the London Mathematical Society* **22** (1890–91), 263.

124. Errera's example appeared in his doctoral thesis, published as *Du Coloriage des cartes et de quelques questions d'analysis situs*, Falk Fils, Brussels and Gauthier-Villars, Paris, 1921; it is discussed by Joan Hutchinson and Stan Wagon in 'Kempe revisited', *American Mathematical Monthly* **105** (February 1998), 170–74. De la Vallée Poussin's example appeared in 'Deuxième réponse à Question 51', *L'Intermédiaire des mathématiciens* **3** (1896), 179–80.

131. For the general solution of the empire problem, see Brad Jackson and Gerhard Ringel, 'Heawood's empire problem, *Journal of Combinatorial Theory B* **38** (1985), 168–78.

138. L. Heffter's paper is 'Ueber das Problem der Nachbargebiete', *Mathematische Annalen* **38** (1891), 477–508; an English translation of part of it appears in *BLW*, pp. 118–23.

For Lucas's *Récréations mathématiques*, see p. 96 above.

The Paris papers were by P. Mansion, 'Question 51'; H. Delannoy and A. S. Ramsey, 'Réponse à Question 51'; C. de la Vallée Poussin, 'Deuxième réponse'; and H. Delannoy, 'Troisième réponse', in *L'Intermédiaire des Mathématiciens* **1** (1894), 20, 192 and **3** (1896), 179–80, 225.

139. Julius Petersen's two notes appeared in *L'Intermédiaire des mathématiciens* **5** (1998), 225–7 and **6** (1899), 36–8.

P. J. Heawood's second paper, 'On the four-colour map theorem', appeared in the *Quarterly Journal of Pure and Applied Mathematics* **29** (1898), 270–85.

141. Heawood's probability arguments appeared in 'Failures in congruences connected with the four-colour map theorem', *Proceedings of the London Mathematical Society (2)* **40** (1936), 189–202.

8 Crossing the Atlantic

Biographies of George David Birkhoff and Oswald Veblen appear in the *DSB*. A biography of Birkhoff appears in Marston Morse, 'George David Birkhoff and his mathematical work', *Bulletin of the American Mathematical Society* **52** (1946), 357–91, and a brief biography appears in Chapter 1 of *FF*.

143. The Minkowski story appears in Constance Reid's *Hilbert*, Springer (1970), 92–3.

147. The information about Wernicke appears in P. J. Federico's manuscript (see Further Reading above). For a brief report of Wernicke's Toronto lecture, see the *Bulletin of the American Mathematical Society* **4** (1897–8), 2, 5.

Wernicke's main paper is 'Über den kartographischen Vierfarbensatz', *Mathematische Annalen* **58** (1904), 413–26.

148. Heesch's 'method of discharging' appeared in his *Untersuchungen zum Vierfarbenproblem*, B. I. Hochschulskripten, 810/810a/ 810b, Bibliographisches Institut, Mannheim-Wien-Zürich, 1969.

151. Many of the results in Philip Franklin's doctoral thesis, 'On the map color problem', Princeton University, 1921, appeared in his paper 'The four color problem', *American Journal of Mathematics* **44** (3) (July 1922), 225–36; this paper appears in full in *BLW*, pp. 171–80, and he later wrote two introductory articles, 'The four color problem', *Scripta Mathematica* **6** (1939), 149–56, 197–210. A useful survey paper from the 1920s is H. R. Brahana, 'The four-color problem', *American Mathematical Monthly* **30** (July/August 1923), 234–43.

Henri Lebesgue's paper was 'Quelques conséquences simple de la formule d'Euler', *Journal de mathématiques pures et appliquées* **9** (1940), 27–43.

153. Veblen's paper was 'An application of modular equations in analysis situs', *Annals of Mathematics (2)* **14** (1912–3), 86–94; it appears in full in *BLW*, pp. 160–66.

Birkhoff's pioneering paper was 'The reducibility of maps', *American Journal of Mathematics* **35** (1913), 115–28.

157. Arthur Bernhart's paper was 'Six-rings in minimal five-color maps', *American Journal of Mathematics* **69** (1947), 391–412. The honeymoon anecdote is from a personal communication by Garrett Birkhoff.

158. The remark about the Kohinoor diamond appears on page 156 of *FF*. For the theses of Franklin and Errera, see pp. 103 and 121 above.

159. For the contributions of Reynolds, Franklin and Winn, see C. N. Reynolds, 'On the problem of coloring maps in four colors I, II', *Annals of Mathematics (2)* **28** (1926–7), 1–15 , 477–92; P. Franklin, 'Note on the four color problem', *Journal of Mathematics and Physics* **16** (1938), 172–84; and C. E. Winn, 'On the minimum number of polygons in an irreducible map', *American Journal of Mathematics* **62** (1940), 406–16.

The work of Paul Valéry is mentioned on p. 32 of *FF* and is described in J. Mayer, 'Une page mathématique de Valéry: le problème du coloriage des cartes', *Bulletin études valéryennes* **25** (1980), 31–43.

163. For Heesch's classification of configurations, see his book cited on p. 109 above.

164. Birkhoff's first paper on chromatic polynomials was 'A determinant formula for the number of ways of coloring a map', *Annals of Mathematics (2)* **14** (1912–3), 42–6; part of this paper appears in *BLW*, pp. 167–9. Birkhoff's paper with D. C. Lewis is 'Chromatic polynomials', *Transactions of the American Mathematical Society* **60** (1946), 355–451.

165. Hassler Whitney's contribution to chromatic polynomials can be found in 'A logical expansion in mathematics', *Bulletin of the American Mathematical Society* **38** (1932), 572–9; part of this paper appears in *BLW*, pp. 181–4. Whitney went on to write some fundamental papers in the related area of graph theory, before turning his attention to the field of algebraic topology, in which he became one of the leading exponents of his generation.

167. The connections between chromatic polynomials and the golden ratio are outlined in G. Berman and W. T. Tutte, 'The golden root of a chromatic polynomial', *Journal of Combinatorial Theory* **6** (1969), 301–2; and explored in W. T. Tutte, 'On chromatic polynomials and the golden ratio', *Journal of Combinatorial Theory* **9** (1970), 289–96.

9 *A new dawn breaks*

For a biography of Heinrich Heesch, see Hans-Günther Bigalke, *Heinrich Heesch: Kristallgeometrie, Parkettierungen, Vierfarbenforschung*, Birkhäuser, Basel, 1988. Brief biographies of Heesch, Haken and Dürre appear in Chapter 1 of *FF*. The quotations by Wolfgang Haken in this chapter are taken from an interview conducted by Tony Dale on 16 April 1994 in Urbana, Illinois, for Donald MacKenzie's 'Slaying the Kraken' (see Further Reading above).

For the papers by Wernicke, Franklin and Lebesgue, see pp. 147 and 151 above.

169. Dénes König's book is *Theorie der endlichen und unendlichen Graphen*, Akademische Verlagsgesellschaft, Leipzig, 1936; an English edition, *Theory of Finite and Infinite Graphs*, was published by Birkhäuser in 1990.

170. The graph theory texts by Berge, Ore, Busacker and Saaty, and Harary are: C. Berge, *Theory of Graphs and Its Applications*, Wiley, 1961; O. Ore, *Theory of Graphs*, American Mathematical Society, 1961; R. G. Busacker and T. L. Saaty, *Finite Graphs and Networks*, McGraw-Hill, 1965; and F. Harary, *Graph Theory*, Addison-Wesley, 1969.

For details of Oystein Ore's book *The Four-Color Problem*, see Further Reading above. The paper by O. Ore and J. Stemple is 'Numerical calculations on the four-color problem', *Journal of Combinatorial Theory* **8** (1970), 65–78.

A full discussion of the solution of the Heawood conjecture appears in G. Ringel, *Map Color Theorem*, Springer, 1974. Further information can be found in G. Ringel and J. W. T. Youngs, 'Solution of the Heawood map-coloring problem', *Proceedings of the National Academy of Sciences, U.S.A.* **60** (1968), 438–45.

174. David Hilbert's lecture, 'Sur les problèmes futurs des mathématiques', appeared in the *Proceedings of the Second International Congress of Mathematicians, Paris* (1902), 58–114. There is a good account of his problems in J. J. Gray, *The Hilbert Challenge*, Oxford University Press, 2000. Heesch's work on tilings is discussed in detail in Bigalke's book, cited above.

177. Wolfgang Haken's Amsterdam lecture on the knot problem is mentioned briefly in the *Proceedings of the International Congress of Mathematicians*, Amsterdam, 1954, and his solution appears in 'Theorie der Normalflächen', *Acta Mathematica* **105** (1961), 245–375.

178. For Kurt Gödel's 1931 paper in English, 'On formally undecidable propositions of the *Principia Mathematica*', see *From Frege to Gödel* (ed. J. van Heijenoort), Harvard University Press, 1967, 596–616.

179. For Birkhoff's paper, see p. 153 above; for Heesch's classification of configurations, see p. 148 above.

183. Edward F. Moore's map appears in L. Steen, 'Solution of the four color problem', *Mathematics Magazine* **49** (4) (September 1976), 219–22; the map that appears in *Scientific American* **237** No. 4 (October 1977), 109, is incorrect.

184. For details of Oystein Ore's book on the four-colour problem, see Further Reading above.

188. Yoshio Shimamoto's remark appears on p. 212 of Bigalke's book *Heinrich Heesch* (see above).

The paper by Hassler Whitney and Bill Tutte is 'Kempe chains and the four-colour problem', *Utilitas Mathematicae* **2** (1972), 241–81; it was reprinted in *Studies in Graph Theory II* (ed. D. R. Fulkerson), Mathematical Association of America (1975), 378–413.

10 Success . . .

Brief biographies of Wolfgang Haken, Kenneth Appel and John Koch appear in Chapter 1 of *FF*. For a biography of Heinrich Heesch, see Hans-Günther Bigalke, *Heinrich Heesch: Kristallgeometrie, Parkettierungen, Vierfar-benforschung*, Birkhäuser, Basel, 1988. The quotations by Haken, Appel and Koch in this chapter are taken from interviews conducted by Tony Dale on 16 April, 6 June and 3 May 1994, for Donald MacKenzie's 'Slaying the Kraken' (see Further Reading above).

190. Some of Heesch's 8900 configurations appear on p. 203 of Bigalke's book *Heinrich Heesch* (see above).

193. Walter Stromquist's paper proving the four-colour theorem for all maps with up to 51 countries is 'The four-color theorem for small maps', *Journal of Combinatorial Theory B* **19** (1975), 256–68. His doctoral thesis was 'Some aspects of the four color problem', Harvard University, 1975.

194. Shimamoto's remark appears on p. 224 of Bigalke's book *Heinrich Heesch* (see above).

198. Appel and Haken's paper on 'The existence of unavoidable sets of geographically good configurations' appeared in the *Illinois Journal of Mathematics* **20** (1976), 218–97. For their paper on maps with no adjacent pentagons, 'An unavoidable set of configurations in planar triangula-tions', see the *Journal of Combinatorial Theory B* **26** (1979), 1–21.

200. Koch's doctoral thesis was 'Computation of four color irreducibility', University of Illinois at Urbana-Champaign, 1976.

205. Allaire and Swart's joint paper, 'A systematic approach to the determination of reducible configurations in the four-color conjecture', appeared in the *Journal of Combinatorial Theory B* **25** (1978), 339–62.
207. After the Appel–Haken solution appeared, Frank Bernhart wrote 'A digest of the four color theorem', *Journal of Graph Theory* **1** (1977), 207–25.
208. Tutte's earlier paper was 'Map coloring problems and chromatic polynomials', *American Scientist* **62** (1974), 702–5.
209. Reports on the Appel–Haken solution appeared in many newspapers and periodicals of 1976, including *The Times* (23 July), *SIAM News* (August), *Science* (13 August), the *Toronto Globe* (24 August), *Le Monde* (1 September), *Time* (20 September), the *New York Times* (24 and 26 September) *Scientific American* (October), *New Scientist* (21 October) and *Die Neue Zürcher Zeitung* (24 October).
211. The official research announcement was K. Appel and W. Haken, 'Every planar map is four colorable', *Bulletin of the American Mathematical Society* **82** (1976), 711–12.
213. The solution to the four-colour problem appeared in two parts, together with a microfiche:
K. Appel and W. Haken, 'Every planar map is four colorable, Part I: Discharging', *Illinois Journal of Mathematics* **21** (1977), 429–90; and
K. Appel, W. Haken and J. Koch, 'Every planar map is four colorable, Part II: Reducibility', *Illinois Journal of Mathematics* **21** (1977), 491–567.

. . . but is it a proof?

The quotations by Haken, Appel and Swart in this chapter are taken from interviews conducted by Tony Dale on 16 April, 6 June and 1 June 1994, for Donald MacKenzie's 'Slaying the Kraken' (see Further Reading above).
214. The quotation 'in my view . . .' appears in F. F. Bonsall, 'A down-to-earth view of mathematics', *American Mathematical Monthly* **89** (1982), 8–15; the remark 'God wouldn't let the theorem . . .' was made by Herbert Wilf to Kenneth Appel in mid-1976.
215. Donald Albers' report 'Polite applause for a proof of one of the great conjectures of mathematics: what is a proof today?' appeared in the *Two-Year College Mathematics Journal* **12** (2) (March 1981), 82.
216. The experience of Armin Haken is recalled in the above-mentioned interviews with Appel and Haken.

217. Thomas Tymoczko's article, 'The four-color problem and its philosophical significance', appeared in *The Journal of Philosophy* **76** (2) (February 1979), 57–83. Another article by him was 'Computers, proofs and
mathematicians: a philosophical investigation of the four-color proof',
Mathematics Magazine **53** (3) (May 1980), 131–8.

218. Ted Swart's response was 'The philosophical implications of the four-
color problem', *American Mathematical Monthly* **87** (November 1980),
697–707.

219. The group theory articles referred to are W. Feit and J. G. Thompson,
'Solvability of groups of odd order', *Pacific Journal of Mathematics* **13**
(1963), 775–1029; and D. Gorenstein, 'The classification of finite simple
groups I', *Bulletin of the American Mathematical Society* **4** (1) (January
1979), 43–200. Andrew Wiles's proof is described in Simon Singh's *Fermat's Last Theorem*, Fourth Estate, London, 1997.

220. Ian Stewart's comment appears on p. 304 of *Concepts of Modern
Mathematics*, Penguin, 1981. Daniel Cohen's remarks appeared in 'The
superfluous paradigm', *The Mathematical Revolution Inspired by Comput-
ing* (ed. J. H. Johnson and M. J. Loomes), Oxford (1991), 323–9. H. S. M.
Coxeter's comments were made in an interview by D. Albers and G. L.
Alexanderson published in *Mathematical People*, Birkhäuser, 1985. George
Spencer-Brown's comments appear in his solution to the four-colour prob-
lem, published as an appendix to the German translation of his *Laws of
Form*, 1997.

222. G. H. Hardy's remark appears in Section 10 of *A Mathematician's
Apology*, Cambridge University Press, 1940.

223. Accounts of G. Spencer-Brown's solution and his lecture appeared in
the *Times Higher Education Supplement*, 17 and 24 December 1976, and in
New Scientist, 23/30 December 1976 and 6 January 1977. His visit to Stanford University is mentioned in Martin Gardner's *The Last Recreations*,
Springer/Copernicus, 1997, 88–9.

224. Frank Allaire's paper 'Another proof of the four colour theorem, I'
appeared in the *Proceedings of the Seventh Manitoba Conference on Numerical Mathematics and Computing* (1977), 3–72.

Appel and Haken's *Scientific American* article and Saaty and Kainen's book
are cited in Further Reading above. T. L. Saaty's article 'Thirteen colorful
variations on Guthrie's four-color conjecture' appeared in the *American
Mathematical Monthly* **79** (January 1972), 2–43.

225. Ulrich Schmidt's dissertation was 'Überprüfung des Beweis für den Vierfarbensatz', Diplomarbeit, Technische Hochschule Aachen, 1982. S. Saeki's observations appear in his *Verification of the Discharging Procedure in the Four Color Theorem*, Master's Thesis, University of Tokyo, 1985.
K. Appel and W. Haken's article 'The four color proof suffices' appeared in the *Mathematical Intelligencer* **8** (1) (1986), 10–20, 58. Their book *Every Planar Map is Four Colorable* was published by the American Mathematical Society in 1989.

226. The recent proof by Neil Robertson, Daniel Sanders, Paul Seymour and Robin Thomas was outlined by Seymour in 'Progress on the four-color theorem', *Proceedings of the International Congress of Mathematicians, Zürich*, Birkhäuser, 1995, and was described by all four of them in 'A new proof of the four-colour theorem', *Electronic Research Announcements of the American Mathematical Society* **2** (1) (August 1996), 17–25, and in 'The four-colour theorem', *Journal of Combinatorial Theory B* **70** (1997), 2–44.
Another outline of the recent proof, with directions for future progress, appeared in Robin Thomas, 'An update on the four-color theorem', *Notices of the American Mathematical Society* **45** (7) (August 1998), 848–59.

227. W. T. Tutte's article 'Colouring problems' appeared in the *Mathematical Intelligencer* **1** (1978), 72–5.

228. It is very appropriate that the final words of this book are by Bill Tutte. Not only was he one of the greatest graph-theorists of the twentieth century, but the sad news of his death reached me on the day I handed the manuscript of this book to the publisher.

Chronology of events

1750 On 14 November, Leonhard Euler states the polyhedron formula in a letter to Christian Goldbach, but is unable to prove it.

1794 Adrien-Marie Legendre gives the first correct proof of the polyhedron formula.

1811—13 Simon-Antoine-Jean Lhuilier obtains a version of the polyhedron formula for a polyhedron with tunnels, and Augustin-Louis Cauchy proves the formula for projections of polyhedra onto the plane.

c. **1840** August Ferdinand Möbius poses the problem of the five princes to his class.

1852 Francis Guthrie finds that four colours suffice to colour a map of England.

On 23 October Augustus De Morgan writes about the four-colour problem to Sir William Rowan Hamilton.

1853/4 De Morgan writes about the problem to William Whewell and Robert Ellis.

1855 Thomas Penyngton Kirkman investigates cycles on polyhedra.

1856 Sir William Rowan Hamilton outlines his icosian calculus and relates it to cycles on a dodecahedron.

1860 On 14 April the four-colour problem appears in print for the first time, in a book review by De Morgan in the *Athenaeum*.

c. **1868** Charles Sanders Peirce presents a solution of the four-colour problem at Harvard University, USA.

1878 On 13 June Arthur Cayley asks about the four-colour problem at a meeting of the London Mathematical Society.

1879 Arthur Cayley writes a note showing that when trying to solve the four-colour problem it is sufficient to consider cubic maps.

Alfred Bray Kempe publishes his purported proof of the four-colour theorem in the *American Journal of Mathematics*; Kempe's proof is discussed in scientific gatherings at Johns Hopkins University, Baltimore.

1880 Peter Guthrie Tait shows that the colouring of maps with four colours is equivalent to the colouring of boundary lines with three colours, and conjectures that every cubic polyhedron has a Hamiltonian cycle.

Frederick Guthrie identifies his brother Francis as the originator of the four-colour problem.

1885 In Leipzig Richard Baltzer confuses the four-colour problem with the problem of the five princes.

1886 The four-colour problem is posed as a challenge problem for the pupils of Clifton College.

1889 Frederick Temple, Bishop of London, confuses the four-colour problem with the problem of the five princes.

1890 Percy Heawood points out the error in Kempe's proof, proves the five-colour theorem, and tries to extend the four-colour problem to empires and to maps on toruses. He proves that every map on the torus can be coloured with seven colours, and produces a torus map that needs seven colours. For a torus with two or more holes, he obtains the correct formula for the number of colours, but fails to prove that there are maps that need this number of colours.

1891 Lothar Heffter points out the error in Heawood's argument for a torus with two or more holes.

1898 Heawood produces a second paper in which he extends Tait's ideas on colouring the boundary lines of a map.

1904 Paul Wernicke produces an unavoidable set of configurations.

1912 George Birkhoff introduces the idea of a chromatic polynomial.

1913 Birkhoff pioneers the study of reducible configurations, proving in particular that the 'Birkhoff diamond' is reducible.

1920 Philip Franklin obtains some new unavoidable sets of configurations and proves that four colours suffice for all maps with up to 25 countries.

1926 Clarence Reynolds extends Franklin's result to maps with up to 27 countries.

1930–32 George Birkhoff and Hassler Whitney obtain further results on chromatic polynomials.

c. **1935** Heinrich Heesch becomes interested in the four-colour problem.

1938 Franklin proves that four colours suffice for all maps with up to 31 countries.

1940 C. E. Winn extends Franklin's result to all maps with up to 35 countries.

Henri Lebesgue obtains some new unavoidable sets of configurations.

1946 George Birkhoff and D. C. Lewis write a lengthy paper on chromatic polynomials.

Bill Tutte produces a cubic polyhedron that has no Hamiltonian cycle, thereby disproving Tait's conjecture of 1880.

c. **1948** Heinrich Heesch proposes to search for an unavoidable set of reducible configurations.

1960s Edward F. Moore shows that any unavoidable set of reducible configurations must necessarily be complicated, containing in particular at least one configuration of ring-size 12 or greater.

c. **1965** Heinrich Heesch and Karl Dürre use a computer to test the reducibility of configurations.

1967 Oystein Ore publishes the first book on the four-colour problem.

1968 Oystein Ore and Joel Stemple prove that four colours suffice for every map with up to 40 countries.

Gerhard Ringel and Ted Youngs complete the proof of the Heawood conjecture on the colouring of maps on a torus with at least two holes.

1969 Heinrich Heesch publishes his book on the four-colour problem, which includes the first discussion of the method of discharging and introduces the terms *D*-reducible and *C*-reducible.

c. **1970** Heinrich Heesch and Wolfgang Haken collaborate on the four-colour problem.

1971 Heesch produces his three obstacles to reducibility.

Yoshio Shimamoto produces his horseshoe configuration, which subsequently turns out not to be *D*-reducible.

1972 Kenneth Appel begins his collaboration with Wolfgang Haken.

1974 John Koch joins Appel and Haken.

1975 Martin Gardner produces his April Fool's column in *Scientific American* on a purported counter-example to the four-colour problem.

1976 On 22 July Kenneth Appel and Wolfgang Haken publicly announce their proof of the four-colour theorem, based on the construction of an unavoidable set of 1936 reducible configurations.

1977 Kenneth Appel, Wolfgang Haken and John Koch publish their proof of the four-colour theorem in the *Illinois Journal of Mathematics*, based on the construction of an unavoidable set of 1482 reducible configurations. T. L. Saaty and Paul Kainen publish their book on the four-colour problem.

1979 Thomas Tymoczko publishes a philosophical paper criticizing Appel and Haken's proof of the four-colour theorem.

1981 Ulrich Schmidt discovers an error in the Appel–Haken proof, which is subsequently corrected.

1984 Brad Jackson and Gerhard Ringel solve Heawood's empire problem in general.

1986 Appel and Haken publish a paper describing their methods and refuting the persistent rumours about their proof.

1989 Appel and Haken produce an extended version of their proof in book form: *Every Planar Map is Four Colorable*.

1994 Neil Robertson, Daniel Sanders, Paul Seymour and Robin Thomas obtain a revised proof of the four-colour theorem. Following Appel and Haken's general approach, they produce an unavoidable set with 633 reducible configurations, using a computer for both parts of the proof.

Glossary

All terms in **bold** can be looked up in the glossary.

Archimedean polyhedron see **semi-regular polyhedron**

Birkhoff diamond A configuration of four adjoining pentagons surrounded by a ring of six countries.

boundary line Part of the boundary of a country, connecting two adjacent meeting points.

buckyball A polyhedral molecule with pentagonal and hexagonal faces.

chromatic polynomial The number of ways of colouring a map with a given number of colours. If there are λ colours, then the number of ways is a **polynomial** in λ.

closed cycle A sequence of lines on a diagram that pass only once through any point and return to the starting point.

configuration A connected collection of countries in a map.

counting formula If C_k is the number of k-sided countries in a **cubic map**, then $4c_2 + 3c_3 + 2c_4 + c_5 - c_7 - 2c_8 - 3c_9 - 4c_{10} - \ldots = 12$.

country A region in a map.

C-reducible configuration A configuration that can be proved to be **reducible** only after it has been modified in some way.

cube A regular polyhedron bounded by six squares.

cubic map A map in which there are exactly three boundary lines at each meeting point.

cubic polyhedron A polyhedron in which exactly three faces meet at each vertex.

cuboctahedron A semi-regular polyhedron with eight triangular and six square faces.

cycle see **closed cycle**.

digon A two-sided country in a map.

discharging see **method of discharging**.

dodecahedron A regular polyhedron bounded by twelve regular pentagons.

D-**reducible configuration** A configuration for which every colouring of the surrounding ring is a **good colouring**, or can be converted into one by one or more **Kempe-chain changes** of colour.

edge (of polyhedron) A line along which two faces of a polyhedron meet.

empire problem The problem of colouring a map with several empires, each consisting of a 'mother country', and a number of 'colonies' that must be coloured the same as the mother country.

Euler's formula Short for **Euler's polyhedron formula**.

Euler's formula for the *h*-holed torus For any map drawn on an *h*-holed torus,

(number of countries) − (number of boundary lines)
+ (number of meeting points) = 2 − 2*h*.

Euler's formula for maps on the plane or sphere For any map drawn on the plane or sphere,

(number of countries) − (number of boundary lines)
+ (number of meeting points) = 2.

Euler's formula for the torus For any map drawn on the **torus**,

(number of countries) − (number of boundary lines)
+ (number of meeting points) = 0.

Euler's polyhedron formula For any polyhedron,

(number of faces) − (number of edges)
+ (number of vertices) = 2.

exterior region The infinite part of the plane lying outside the main part of a map.

face (of polyhedron) One of the flat surfaces that make up a polyhedron.

finitization of the four-colour problem Reducing the proof of the four-colour problem to investigating only a finite number of configurations.

five-colour theorem Every map can be coloured with at most five colours in such a way that neighbouring countries are coloured differently.

four-colour problem Can every map drawn on the plane be coloured with at most four colours in such a way that neighbouring countries are coloured differently?

four-colour problem for a sphere Can every map drawn on the surface of a sphere be coloured with at most four colours in such a way that neighbouring countries are coloured differently?

four-colour theorem Every map drawn on the plane can be coloured with at most four colours in such a way that neighbouring countries are coloured differently.

four colours suffice A statement of the four-colour theorem.

four-legger country One of Heesch's three **obstacles to reducibility**.

geographically good configuration A configuration that contains neither of Heesch's first two **obstacles to reducibility**.

golden ratio The number $\frac{1}{2}(1 + \sqrt{5}) = 1.618034. \ldots$

good colouring A colouring of a ring of countries that can be extended directly to a colouring of the countries inside the ring.

graph see **linkage**.

great rhombicuboctahedron A semi-regular polyhedron with twelve square, eight hexagonal and six octagonal faces.

Guthrie's problem Another name for the four-colour problem.

Hamiltonian cycle A sequence of lines on a diagram that pass exactly once through every point and return to the starting point.

hanging 5–5 pair One of Heesch's three **obstacles to reducibility**.

Heawood conjecture For each positive number h, there is a map on the surface of an *h***-holed torus** that requires $H(h) = \frac{1}{2}(7 + \sqrt{1 + 48h})$ colours; this conjecture was proved by Ringel and Youngs.

Heawood number The number $H(h) = \frac{1}{2}(7 + \sqrt{1 + 48h})$ associated with the Heawood conjecture.

heptagon A seven-sided country in a map.

hexagon A six-sided country in a map.

h-**holed torus** A ring-shaped surface with h holes.

horseshoe see **Shimamoto horseshoe**.

icosahedron A regular polyhedron bounded by twenty equilateral triangles.

icosian calculus An algebraic system of symbols that satisfy certain equations; it was introduced by Hamilton.

Icosian Game A game in which Hamiltonian cycles are traced on a dodecahedron.

induction see **mathematical induction**.

Kempe chain Part of a coloured map consisting of countries coloured with just two colours.

Kempe-chain argument see **method of Kempe chains**.

Kempe-chain change of colour An interchange of two colours in part of a coloured map.

knot problem The problem of determining whether a given tangle of string in three dimensions contains a knot.

kraken A fabled Norwegian sea monster.

linkage or **graph** A diagram obtained from a map, with points representing countries and linked points corresponding to neighbouring countries. Colouring the map corresponds to lettering the points so that any points that are linked are lettered differently.

***m*-and-*n* rule** A useful rule of thumb for identifying reducible configurations.

map A collection of countries or regions separated by boundary lines.

mathematical induction A method of mathematical proof. As applied to maps, one derives the truth of a statement for maps with $n + 1$ countries from the corresponding statement for maps with n countries, and then deduces the result for all maps.

meeting point A point in a map where boundary lines and countries meet.

method of discharging A method for proving that a set of configurations is an **unavoidable set**: each k-sided country is assigned a 'charge' of $6 - k$, and the charges are then moved around the map so that the total charge remains unchanged.

method of Kempe chains A method of colouring maps in which two colours are interchanged to enable the colouring of countries that could not previously be coloured.

minimal counter-example see **minimal criminal**.

minimal criminal A map with a certain number of countries that cannot be coloured with four (or any other given number of) colours, while any map with fewer countries can be so coloured.

neighbouring countries or **regions** Two countries (or regions) with a boundary line in common.

nonagon A nine-sided country in a map.

obstacles to reducibility Three arrangements of countries (a **three-legger articulation country**, a **four-legger country** and a **hanging 5–5 pair**) whose appearance in a configuration seems to prevent it from being a **reducible configuration**.

octagon An eight-sided country in a map.

octahedron A regular polyhedron bounded by eight equilateral triangles.

'only five neighbours' theorem Every map has at least one country with five or fewer neighbours.

'only six neighbours' theorem for the torus Every map on the torus has at least one country with six or fewer neighbours.

pentagon A five-sided country in a map.

Petersen graph A diagram with ten points and fifteen lines that has no **Hamiltonian cycle** and does not arise from a polyhedron.

Platonic solid see **regular polyhedron**.

Poincaré conjecture A problem concerning spheres in four-dimensional space.

polyhedron A solid shape bounded by plane faces.

polynomial An expression, such as $\lambda^4 - 5\lambda^3 + 8\lambda^2 - 4\lambda$, consisting of multiples of powers of a variable, in this case λ.

problem of the five palaces Five sons each build a palace. How can the five palaces be linked in pairs by roads so that no two roads cross?

problem of the five princes A king in India had a large kingdom and five sons. How can his kingdom be divided into five regions so that each region has a boundary line in common with the remaining four regions?

qualitative approach In the case of the four-colour theorem, an attempt to prove it by showing that all maps of a certain type can be coloured with four colours.

quantitative approach In the case of the four-colour theorem, an attempt to prove it by finding the number of ways in which a map can be coloured with any given number of colours.

reducible configuration A configuration that cannot occur in a **minimal criminal**. If a map contains a reducible configuration, then any colouring of the rest of the map with four colours can be extended (possibly after some recolouring) to a colouring of the entire map.

reduction obstacles see **obstacles to reducibility**.

region A general term for a country, county or state in a map.

regular polyhedron or **Platonic solid** A polyhedron in which all faces are regular polygons of the same type, and each vertex has the same arrangement of polygons around it. There are only five regular polyhedra – the tetrahedron, cube, octahedron, dodecahedron and icosahedron.

ring of countries The countries surrounding a given configuration. If there are *k* countries in the ring, the configuration is called a *k*-ring configuration.

ring-size The number of countries surrounding a configuration. If there are *k* surrounding countries, the configuration has ring-size *k*.

semi-regular polyhedron or **Archimedean polyhedron** A polyhedron in which all the faces are regular polygons, not all of the same type, and each vertex has the same arrangement of polygons around it.

seven-colour theorem for the torus Every map drawn on the torus can be coloured with at most seven colours, and there are torus maps that need all seven colours.

Shimamoto horseshoe A configuration with **ring-size** 14 whose **D-reducibility** would have implied the four-colour theorem.

six-colour theorem Every map can be coloured with at most six colours in such a way that neighbouring countries are coloured differently.

square A four-sided country in a map.

stereographic projection A projection, from the 'North Pole', of a map of the surface of a sphere onto the plane on which the sphere sits.

tetrahedron A regular polyhedron bounded by four equilateral triangles.

three-legger articulation country One of Heesch's three **obstacles to reducibility**.

torus A ring-shaped surface with a hole.

triangle A three-sided country in a map.

truncated icosahedron A semi-regular polyhedron with twelve pentagonal and twenty hexagonal faces.

truncated octahedron A semi-regular polyhedron with six square and eight hexagonal faces.

unavoidable set of configurations A collection of configurations, at least one of which must appear in every map. The simplest example consists of a digon, a triangle, a square and a pentagon.

vertex (of polyhedron) A corner of a polyhedron.

Picture Credits

p. 15 copyright © April 1975 by *Scientific American*, Inc. All rights reserved;
p. 17: from TCD MS1493/668, reproduced by courtesy of The Board of
Trinity College, Dublin; pp. 22, 78: courtesy of the London Mathematical
Society; p. 29: August Ferdinand Möbius, *Gesammelte Werke*, Hirzel, Stutt-
gart, 1885; pp. ii, 39, 74, 108, 111, 117: collection of the author; p. 45: P.-H.
Fuss, *Correspondence mathematique et physique de quelques célèbres
géomères du XVIIIème siècle*, St. Petersburg, 1843; p. 61: *Illustrated London
News*, 15 September 1883; p. 88: reproduced from the *American Journal of
Mathematics* courtesy of the Johns Hopkins Press; p. 100: C. Knott, *Life and
Scientific Work of Peter Guthrie Tait*, Cambridge University Press, 1911;
p. 120: reproduced from *The Quarterly Journal of Pure and Applied Mathe-
matics* with permission from Oxford University Press; p. 154: from G. D. Bir-
khoff, *Collected Mathematical Papers* courtesy of the American
Mathematical Society and Dover Publications; p. 171: courtesy of Gerhard
Ringel; pp. 173, 175, 179, 185, 187: from H.-G. Bigalke, *Heinrich Heesch: Kristallgeometrie, Parkettierungen, Vierfarbenforschung*, Birkhauser, Basel, 1988,
supplied by H.-G. Bigalke; p. 191: courtesy of the University of Illinois at
Urbana-Champaign; p. 199: reproduced from the *Journal of Combinatorial
Theory* with permission from Academic Press, Inc., Harcourt Publishing
Division; p. 200: courtesy of John Koch; p. 206: reproduced courtesy of
the *Illinois Journal of Mathematics*; p. 201: courtesy of Kenneth Appel.

Index